P9-CFZ-403

the new BIOtechnology

A0002600355933

the new BIOtechnology
Putting Microbes to Work

Cynthia S. Gross

Lerner Publications Company   Minneapolis

WITHDRAWN
KIRTLAND PUBLIC LIBRARY

620.8

To Gary, with love

Copyright © 1988 by Lerner Publications Company

All rights reserved. International copyright secured.
No part of this book may be reproduced in any form whatsoever without permission in writing from the publisher except for the inclusion of brief quotations in an acknowledged review.

Library of Congress Cataloging-in-Publication Data

Gross, Cynthia S.
The new biotechnology: putting microbes to work / Cynthia S. Gross.
p. cm.
Includes index.
Summary: Describes the new field of biotechnology and the potential and dangers involved in its development.
ISBN 0-8225-1583-0 (lib. bdg.)
1. Biotechnology—Juvenile literature. [1. Biotechnology.] I. Title.
TP248.2.G76 1988
660'.6—dc19 88-18823
CIP
AC

Manufactured in the United States of America

1 2 3 4 5 6 7 8 9 10 97 96 95 94 93 92 91 90 89 88

CONTENTS

This peptide synthesizer is used in making monoclonal antibodies, one of the revolutionary products created by the new biotechnology.

THE NEW BIOTECHNOLOGY

Gene splicing. Genetic engineering. Designer genes. Cloning. Recombinant DNA. These terms leap out at us from newspaper headlines almost daily. They refer to an area of modern science that is playing an increasingly important role in all our lives—biotechnology.

People have different ideas about what biotechnology involves, and many of these ideas are wrong. Biotechnology is not the creation of life in the laboratory. Nor is it making copies of people, altering human beings to build a super race, or manufacturing monsters.

In simplest terms, biotechnology is the use of living organisms to make products. In a larger sense, it also includes the development of microorganisms such as bacteria and yeasts to perform particular tasks, as well as the improvement of plant and animal breeds through biological means.

You can see from this definition that biotechnology is not at all new. For perhaps as long as 10,000 years, the human race has enlisted the aid of microorganisms, often without even knowing what they were. People have used yeasts to make their bread rise and have produced cheese, yogurt, and other milk products with the help of bacteria and yeasts. They have employed bacteria in the processing of leather and linen. They have even manipulated genetic material by breeding their finest animals and saving the best seeds to plant for next year's crops.

As humans have learned more about microorganisms, they have discovered more uses for them and for the chemical substances that they make. The development of antibiotics to fight infection and the use of bacteria to create efficient sewage treatment plants are only two examples of the important roles that microorganisms play in modern life.

Much of the work being done today in the field of biotechnology is not so different from earlier uses of natural processes to make things that we need. The aspect of biotechnology that has brought it so prominently into today's headlines, however, is the development of new methods for the manipulation of genetic material. In the past 10 years or so, there has been a rush of progress in the human ability to tailor genetic material for a particular purpose. Before these techniques were worked out, researchers could only select material from what was available in nature.

The rapid progress that has taken place in biotechnology is based on developments in three different areas. They are recombinant DNA technology, cell fusion, and bioprocess engineering.

Recombinant DNA (rDNA) technology, also known as gene splicing, genetic engineering, or gene cloning, involves taking a portion of the genetic material DNA from one organism and putting it into another organism. Recombination of genetic material occurs constantly in nature, but only recently have researchers been able to carry it out in a controlled fashion in the laboratory. Using this technique, it is possible to make the organism receiving the foreign DNA manufacture a substance that it was not able to make before. Gene splicing can turn a bacterial or yeast cell into a tiny, efficient factory producing such substances as human growth hormone.

Another method of manipulating genetic information is **cell fusion**, the merging of two different types of cells into a single cell. This single cell contains the DNA of both the original cells and thus also possesses their desirable qualities. One of the most promising applications of cell fusion technology is the making of **monoclonal antibodies**, which may eventually prove to be an effective treatment for cancer. Monoclonal antibodies are already in use to diagnose and treat certain diseases in animals as well as to detect the presence of some human hormones.

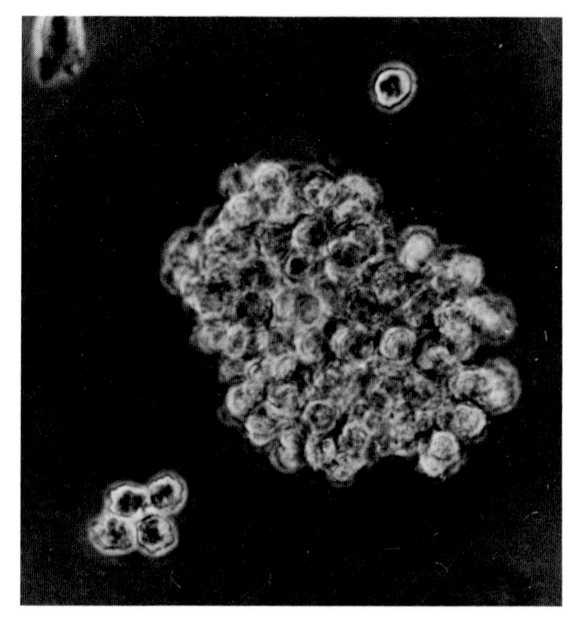

A colony of fused cells producing monoclonal antibodies

The third important factor in the development of the new biotechnology is **bioprocess engineering.** A **bioprocess** is a process that uses living cells or some component of cells to make a product. Using yeast to make bread rise, for example, is a bioprocess. Bioprocess technology consists of planning facilities and maintaining the proper conditions for the efficient and cost-effective use of bioprocesses on an industrial scale. Its chief goal is to adapt biological methods developed in the laboratory for the manufacturing of products in commercial quantities.

Today products made by the new biotechnology are already on the market. They include insulin identical to that produced by the human body, hormones for the treatment of certain deficiencies, enzymes for synthesizing chemicals, and kits for the diagnosis of diseases. This is only the beginning of what promises to be a long line of products created by this amazing new technology.

As with many new and unfamiliar fields of scientific research, biotechnology has been the subject of considerable controversy. Some members of the general public as well as some scientists believe that there are dangers involved in the manipulation of genetic material. They fear that epidemic diseases or ecological disasters might result from using recombinant DNA techniques. Other people are more interested in the business opportunities offered by the new biotechnology. They see the possibility of making big profits by investing in companies on the leading edge of progress in the field.

Some scientists and others involved in research look on biotechnology as a powerful force for improving human life. These people believe that advances in such areas as disease prevention and treatment, food and nutrition, manufacturing, energy production, and pollution control have the potential to benefit all the people of the world, in developed and developing countries alike. From this point of view, biotechnology is seen as offering solutions to some of the most serious problems the human race faces today.

No matter what their point of view, many people foresee that the influence of biotechnology will be far-reaching enough to be considered a revolution on the same scale as the Industrial Revolution of the 19th century. Some even predict that the development of biotechnology could have more impact on society than the introduction of the computer. After reading this book, you will be better prepared to make your own decision about this powerful technology that has already begun to affect all our lives.

Colonies of bacterial cells containing recombinant DNA. The ability to combine genetic material from two different organisms has made the new biotechnology possible.

1

UNRAVELING THE THREAD OF LIFE

WHAT IS DNA?

In order to understand what makes biotechnology possible, it is necessary to know something about DNA and our recently gained ability to manipulate it.

The initials DNA stand for **deoxyribonucleic acid**, a chemical substance that is a vital ingredient in all forms of organic life. The role of DNA is closely related to that of another chemical substance upon which life depends, **protein**. Every cell of every organism contains hundreds or even thousands of different proteins. Some are part of the structure of a cell, such as the surrounding membrane. Many proteins play a role in the work of cells, for example, regulating or speeding up chemical reactions such as using sugars to make energy. Other proteins fight disease-causing invaders or serve as messengers to other cells.

In all living things, the production, or synthesis, of proteins is controlled by DNA. The DNA in the nucleus of each cell directs the synthesis of proteins produced by that cell. In addition to directing the production of proteins, DNA has the ability to pass this vital information on to new cells and even to a new generation of organisms.

The DNA in cells is organized into tiny, threadlike structures called **chromosomes**. Each species of animal and plant has a characteristic number of chromosomes in its body cells. Humans, for example, have 46 chromosomes, grouped into 23 pairs. Mice have 40 chromosomes in 20 pairs, and bacteria have only a single, unpaired chromosome. With the exception of cells involved in reproduction, all the cells in an organism have exactly the same number of chromosomes.

Arranged along the chromosomes are hundreds, even thousands, of regions of DNA known as **genes**. Each gene contains the code for the production of a single specific protein. Because all cells have an identical set of chromosomes, it follows that they all have the same genes. A cell in the human liver, for example, has not only the genes for making proteins needed

THE PARTS OF A BODY CELL

The largest part of the cell is the *cytoplasm*. It is surrounded by a two-layered *cell membrane* that allows substances to pass in and out of the cell.

Inside the cytoplasm are many tiny structures known as *organelles*. Some, like the *mitochondria*, produce energy for the cell's work.

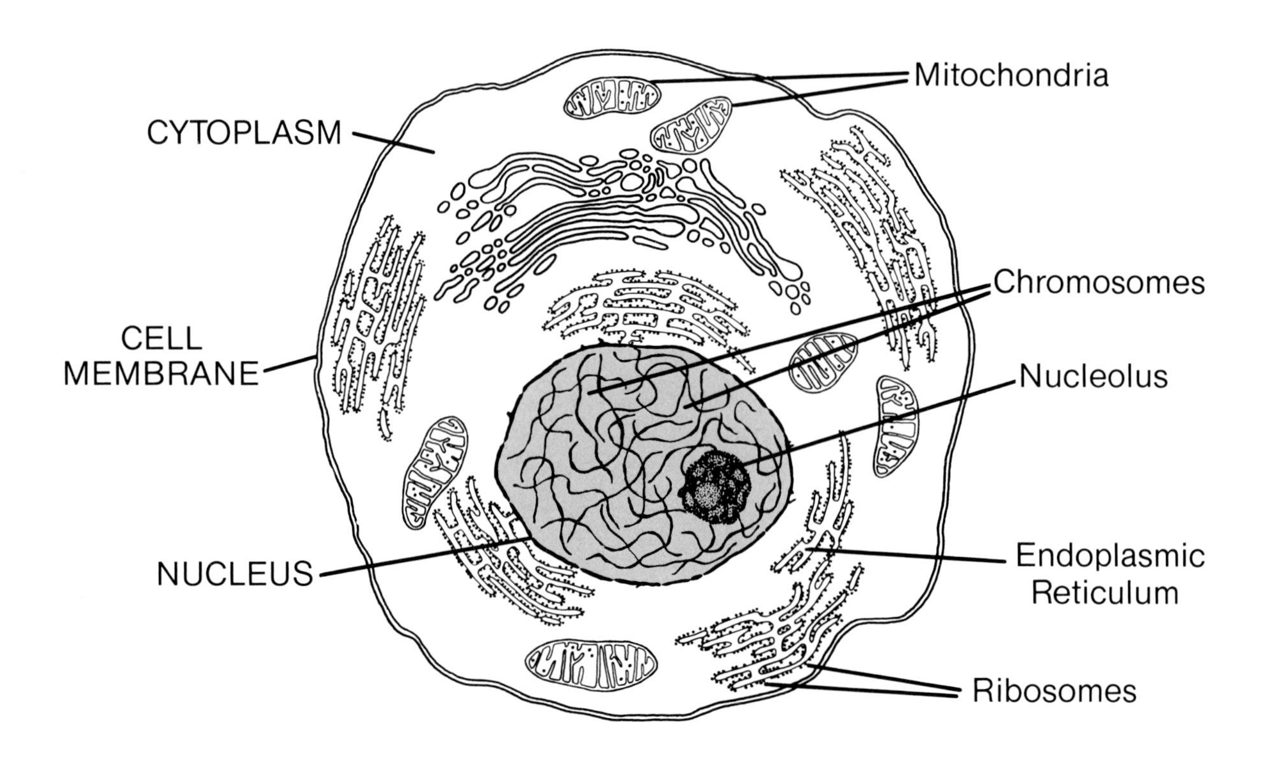

The *nucleus* controls all the functions of a cell. It contains threadlike structures called *chromosomes*, which are made up of DNA. A round body known as the *nucleolus* is also found within the nuclei of many cells.

The *endoplasmic reticulum* is a network of spaces enclosed by membrane. Attached to part of this organelle's surface are round *ribosomes*, "factories" where the production of proteins take place.

for its own structure and function but also the genes for the proteins in toenails. The vast majority of genes in a cell remain switched off most of the time. Specific genes are turned on only when the protein for which they contain the blueprint is needed.

THE STRUCTURE OF DNA

It was not until the 1940s that DNA was recognized as the basic genetic material in all living organisms. After learning the identity of this vital substance, researchers raced one another in an attempt to discover its structure and function. The scientists first to propose the model of DNA that we accept today were the American biologist James Watson and the English biochemist Francis Crick, both of whom did their work at Cambridge University in England.

To determine the structure of DNA, Watson and Crick used the technique of X-ray diffraction, which produces patterns as X-rays are bounced off parts of molecules. They also used model-building to help understand how DNA is constructed. In 1953, the two scientists published their results.

Watson, Crick, and other scientists had learned that DNA molecules are very large, usually the largest in a cell. They are also very long. Each chromosome in a cell seems to consist of a single DNA molecule. An average human chromosome, for example, is made up of one molecule that would be about 2 inches (5 centimeters) long if it were stretched out. Despite their size and their length, DNA molecules are so narrow that they cannot even be seen except with an electron microscope.

The structure of the DNA molecule as described by Watson and Crick is complicated, but a simple image can help to explain it. Picture a ladder, a flexible rope ladder that can be twisted to resemble a spiral staircase. The legs of the ladder are formed by units of two different chemicals that alternate along the entire length of the DNA molecule. One of these chemicals is the sugar deoxyribose, which gives DNA the first part of its full name, deoxyribonucleic acid. The other is phosphate.

The steps or rungs of the DNA ladder extend at right angles from the sugar units of the two legs. Each rung is composed of two chemicals that belong to a group known as **nitrogen bases**. DNA contains four different nitrogen bases: adenine (A), guanine (G), thymine (T), and cytosine (C). The way in which the bases combine to form the rungs of the ladder determines both the structure and the function of DNA.

The nitrogen bases in DNA, like many other types of chemical substances, contain atoms that are joined to form rings. Adenine and guanine, which belong to the group of bases called **purines**, contain two rings. Thymine and cytosine, which are **pyrimidines**, have only one ring.

THE STRUCTURE OF DNA

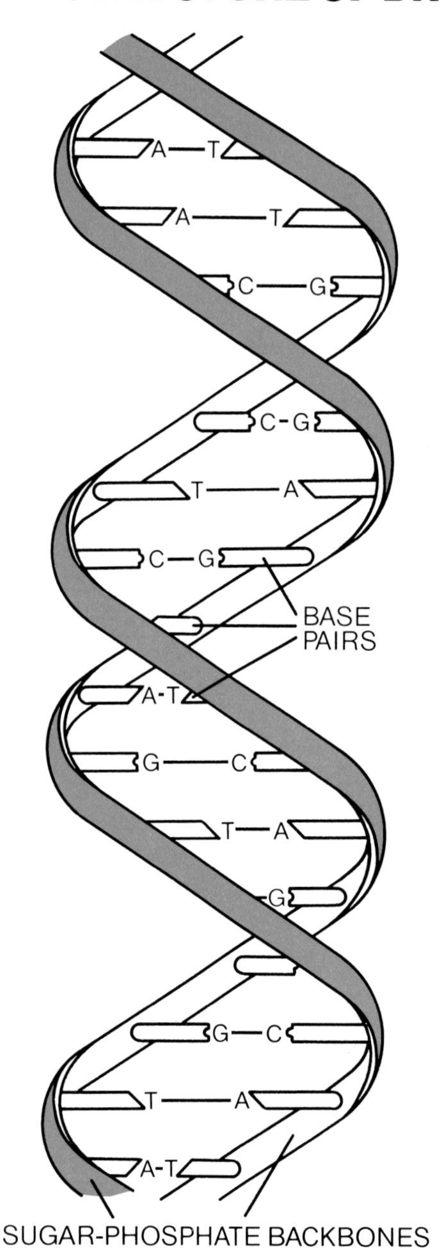

These different structures help to determine which bases will join to form the rungs of the DNA ladder.

The space occupied by the rungs is the same throughout the DNA molecule, and it will accommodate exactly three rings. This means that the only combinations possible are between a purine with two rings and a pyrimidine with one ring. There is not enough space between the ladder sides to permit a purine to be bonded to another purine, making four rings. On the other hand, the space is too wide to be filled by the two rings of two linked pyrimidines.

To get the necessary three rings, however, a pyrimidine cannot link with just any purine. There are other differences in the chemical structures of the bases that determine which members of the two groups will join. The two bases that make up a rung in the DNA ladder are held together by the bonding of certain hydrogen atoms in each base.

The shape of the pyrimidine thymine places two hydrogen atoms in a position for bonding. The purine adenine also has two hydrogen atoms in this kind of position. Cytosine, the other pyrimidine, has three hydrogen atoms available for bonding, as does the second purine, guanine. Therefore, cytosine is always linked to guanine to form a step, and thymine is always joined to adenine. One or the other of these two combinations, known as **complementary pairs**, make up the billions of steps in a DNA molecule.

COMPLEMENTARY BASE PAIRS IN DNA

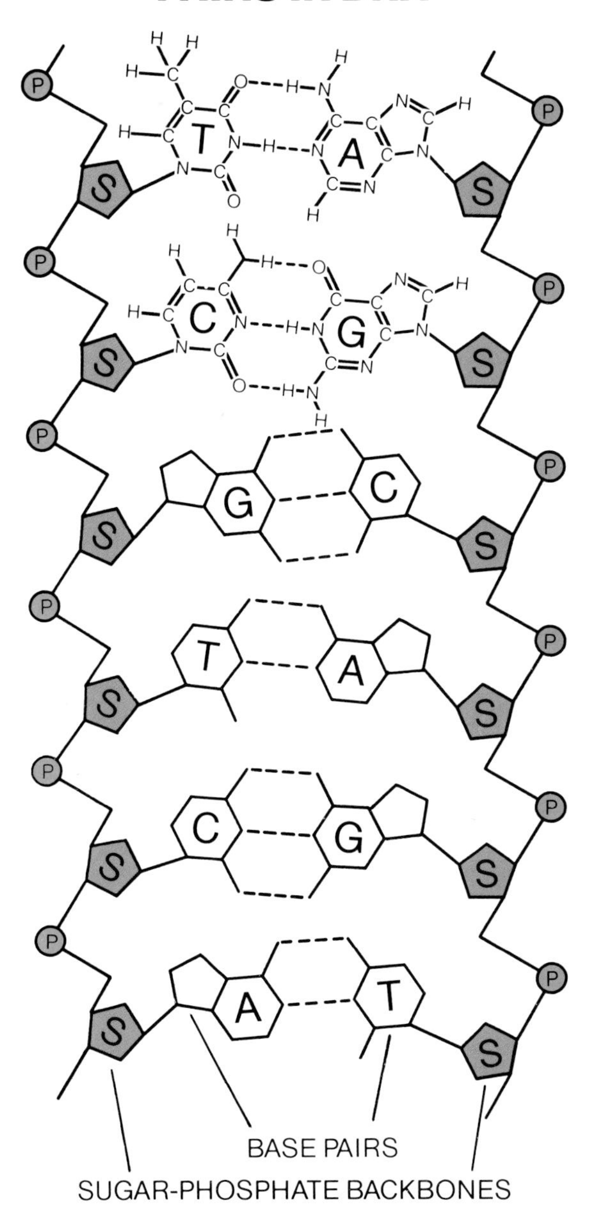

As you study a drawing of DNA, it is very easy to think of the molecule as flat, like the paper on which the drawing is printed. It is not flat; instead, it has three dimensions. Keep in mind that the legs of the DNA ladder are twisted in a spiral shape—the famous double helix discovered by Watson and Crick. The DNA molecule makes a complete turn every 10 steps. This coiling puts the hydrogen atoms of one base in the proper position to bond with those of the other base making up the step.

THE ROLE OF DNA IN THE CELL

DNA plays two essential roles in the life of an organism. One is the production of proteins vital to the functioning of cells. The other is making copies of itself so that the life of the organism can continue over time.

MAKING PROTEINS

The way in which DNA directs the production of proteins is complicated and fascinating. Proteins are large, complex molecules made up of one or more chains of smaller molecules called **amino acids**. There are 20 different kinds of amino acids, and they must be lined up in a precise order to form a specific protein. The order of the amino acids is determined by the sequence or order of the bases in the gene that contains instructions for the production of the protein.

Since there are only four nitrogen bases in DNA, how can the order in which they occur determine the order of 20 different amino acids? The answer is the **codon**, a code of three bases arranged in a specific sequence. Each amino acid is coded for by one (or several) of these groups of bases. For example, the codon guanine-cytosine-adenine (GCA) along a strand of DNA specifies the amino acid alanine. This codon codes for alanine in the cells of any living organism—human, sea slug, bacterium, or geranium.

There are 64 different codons that can be formed from the four bases of the DNA "alphabet." Some of them code for the same amino acid, while others give instructions as to where the amino acid chains should be started and stopped. All the codons together make up the **genetic code**, which is the basic blueprint for all life.

How do the codons in a cell actually direct the formation of amino acid chains to make proteins? Here is a simplified description of what happens. When a particular protein is to be made by the cell, the region of DNA bearing the gene that carries the instructions for the protein unwinds along its length. The sequence of codons in this region serves as a template, or pattern, for making a corresponding section of another vital chemical, RNA. RNA, or **ribonucleic acid**, is much like DNA, with two exceptions. The thymine that occurs in DNA is replaced in RNA by uracil, a similar pyrimidine. The sugar in RNA is ribose instead of deoxyribose.

The single strand of RNA formed from the DNA is called messenger RNA (mRNA) because it carries the genetic code from the cell nucleus into the cytoplasm, the main part of the cell. In the cytoplasm, mRNA goes to a **ribosome**, one of the tiny bodies in the cell that serve as "factories" for protein production. With the aid of the ribosome and another form of RNA, transfer RNA (tRNA), mRNA collects the specified amino acids from the cytoplasm and puts them together in the proper order. The amino acids join into a chain or chains that form the protein coded for by the gene.

The way a protein works depends largely upon its shape, the way in which the chains twist around on themselves. For a protein to have the correct shape, it must contain the proper animo acids in exactly the right order. If the genetic code for the protein is flawed in some way—for example, if one base is out of order—the structure of the protein might be changed, affecting its function. Such genetic "mistakes" can result in disorders like sickle cell anemia, in which certain blood proteins do not function properly.

REPLICATION

The other essential job of DNA is **replication**, or making copies of itself. When a cell divides, the genetic information contained in the DNA must be passed to each of the two daughter cells.

FORMATION OF RNA

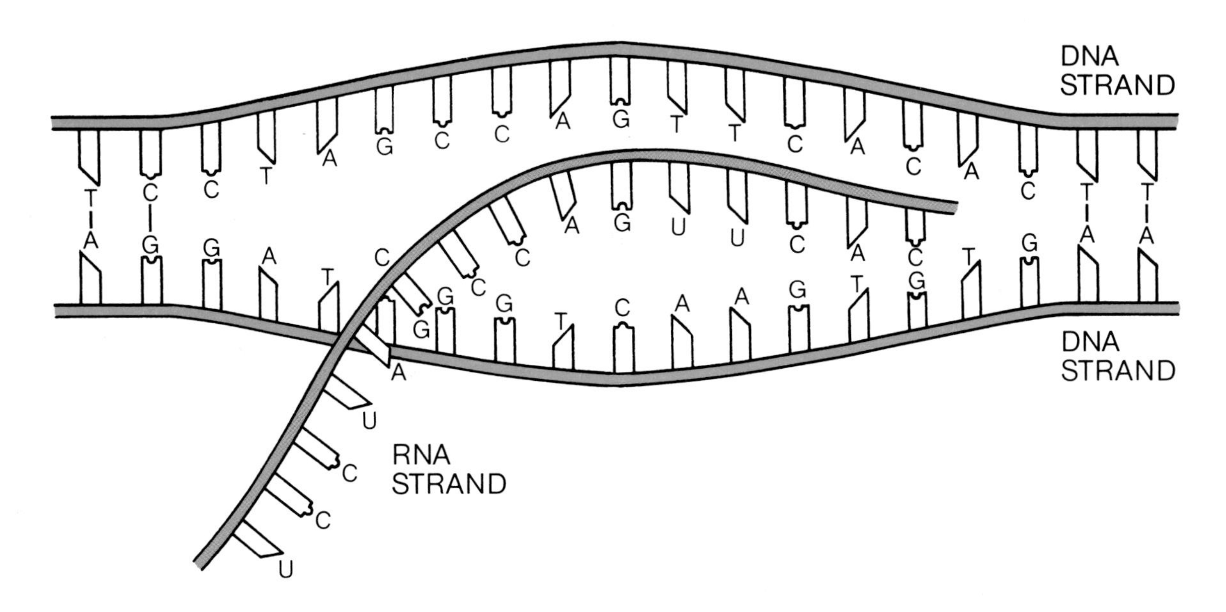

During replication, the DNA ladder of each chromosome comes apart at the hydrogen bonds between the bases that make up the ladder's rungs. New bases from the supply inside the cell's nucleus line up to bond with their complementary bases on each original strand. Attached to each new base is a sugar and phosphate that will form the second strand or leg of the new DNA ladder. The unit of base, sugar, and phosphate is known as a **nucleotide**.

After replication, there are two chromosomes instead of one, each a double helix with a strand from the original DNA molecule plus a newly formed strand. When the cell divides, each daughter cell gets a full set of chromosomes containing exactly the same DNA as in the parent cell. In this way, genetic information is passed on to each succeeding generation of cells.

When human reproductive cells divide, the number of chromosomes in each new cell is 23, half the number in body cells. If a reproductive cell joins with another reproductive cell, the result will be a new cell containing 46 chromosomes. Through this union, the genetic material from two different organisms is joined to form a new individual with characteristics of both parents.

REPLICATION OF DNA

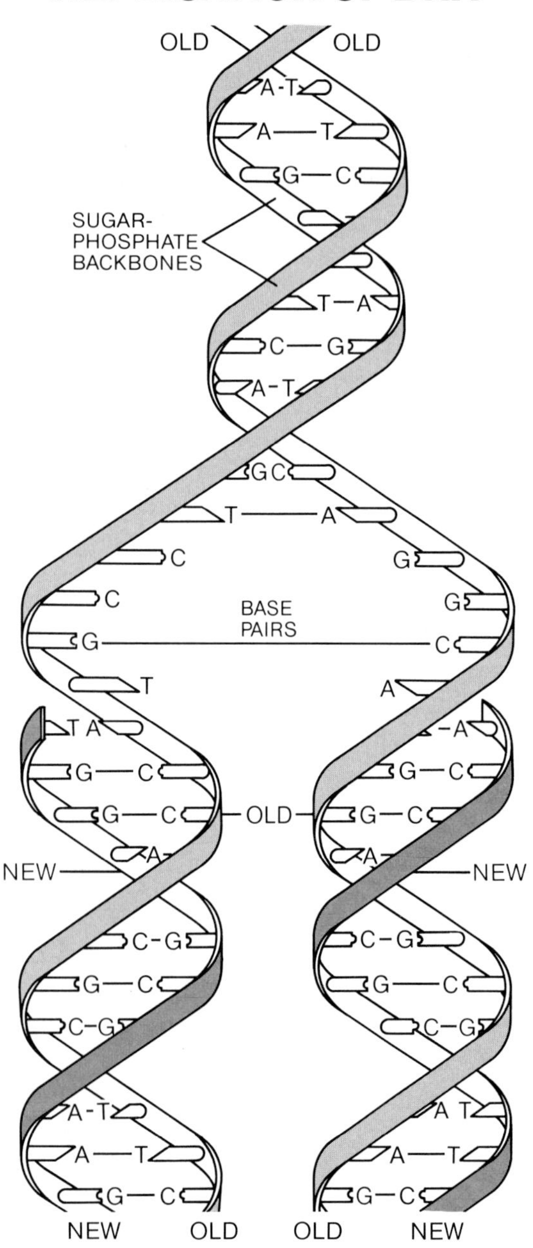

RECOMBINANT DNA

Once the structure of DNA was determined, that knowledge became the basis of research into ways to manipulate genetic material. In 1972, Paul Berg of Stanford University and his students joined DNA fragments from two viruses into molecules of recombinant DNA. One year later, the first successful insertion of a piece of genetic material from one bacterial species into another was performed by Herbert Boyer of the University of California at San Francisco and Stanley Cohen of Stanford University. These two researchers also succeeded in getting their recombined bacteria to make a protein that it was unable to make before. Their experiment laid the groundwork for the development of recombinant DNA technology.

The aim of rDNA technology is to trick a cell into producing some chemical substance that it cannot make naturally and for which it may have no use or need. This is done by changing the composition of the cell's genetic material, slipping in instructions for the desired material, and providing the ideal conditions for the cell to carry them out. The whole process is usually referred to as genetic engineering.

TOOLS OF GENETIC ENGINEERING

The organisms most often used as host cells in genetic engineering are bacteria and yeasts, although work is also being

Herbert Boyer (left) of the University of California at San Francisco and Stanley Cohen (right) of Stanford University were pioneers in the development of recombinant DNA technology.

done with plant and animal cells.

Bacteria (singular, bacterium) are tiny organisms consisting of a single cell. Individual bacterial cells can be seen only with a microscope, which is the reason that bacteria were not known to exist until the invention of the microscope in the 1600s. A typical cell of the type of bacteria most commonly used for genetic engineering is only 0.5 to 1 micron wide and 2 to 3 microns long. (A micron is 1/1000 of a millimeter or about 1/25,000 of an inch.) Yet each of these tiny cells is able to make hundreds of chemical substances, directed by the information contained in its genetic material.

Unlike more complex organisms, the DNA of a bacterial cell is not confined in a nucleus, a structure consisting of genetic material surrounded by a membrane. A bacterium's single chromosome is contained within the cell's cytoplasm. Many bacteria also include **plasmids**, which are rings of DNA separate from the main chromosome. Plasmids often carry genes for resistance to certain antibiotics.

Another characteristic of plasmids is their ability to pass between cells of the same or even different types of bacteria. Because of this, they are responsible for a great deal of the exchange of genetic

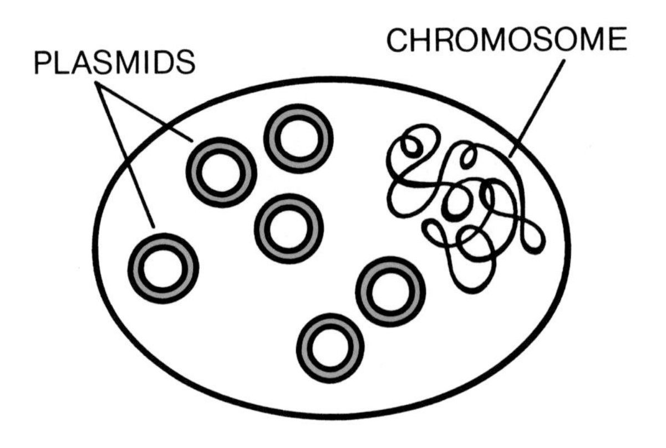

information that occurs in nature. As we shall see, this ability to carry DNA between organisms makes plasmids invaluable in rDNA technology.

Bacteria reproduce by division, each parent cell splitting to form two identical daughter cells. The rate at which bacteria reproduce varies from species to species but, under ideal conditions, may be as rapid as one division every 20 minutes. In theory, a single bacterial cell could produce billions and billions of new cells in less than 24 hours. Fortunately for the other forms of life on earth, this growth is usually limited by conditions such as lack of food, temperatures that are too high or too low, and buildup of waste products.

Bacteria exist everywhere—in the air, in water and soil, in and on our bodies. In nature, many different kinds of bacteria occur together, but for scientific uses, they are grown in pure cultures that contain only one type. The bacterial species used most often in recombinant DNA work is *Escherichia coli*, commonly known as *E. coli*. A normal and usually harmless inhabitant of the human intestinal tract, it has been studied in great detail for many years. More is known about the internal workings and genetic structure of *E. coli* than any other species.

Dozens of different strains of *E. coli* exist, each with small variations in nutritional requirements, structure, and so on. The ones used for rDNA research are weakened strains unable to cause infection. The most common of these is *E. coli* K12. This bacterium is easily cultured, requiring only the sugar glucose and several simple, inexpensive chemicals for nutrition, warmth, and oxygen. Supplied with these materials, *E. coli* divides as often as every 20 minutes.

In addition to bacteria, **yeasts** are also employed as hosts in recombinant DNA work. The type most often chosen is *Saccharomyces cerevisiae*, also known as brewer's yeast, the same organism used for making alcoholic beverages and grown as a food supplement. Yeasts are single-celled microscopic fungi, larger and more complex than bacteria. They have four chromosomes, which, unlike bacterial chromosomes, are contained in a nucleus.

Viruses figure prominently in genetic engineering, some as tools and some as disease-causing agents to be fought. They are much smaller than bacteria and can be seen only with an electron microsope. A viral particle consists of a protein outer coat that surrounds the genetic material, either DNA or RNA. The genetic material of some viruses can be altered by the addition of foreign DNA, which they carry into a host cell.

Other important substances in the rDNA toolbox are **enzymes**. These are protein molecules found in living organisms that act as catalysts, agents necessary for a chemical reaction but not themselves changed by it. Without enzymes, many reactions would take place very slowly or not at all.

It is often possible to recognize a particular chemical compound as an enzyme by its name. The names of enzymes usually (but not always) end in "-ase," with this suffix attached to the name of the substance that is changed in the reaction. Sucrase, for example, is an enzyme that acts upon the sugar sucrose.

Enzymes are invaluable in rDNA technology because they provide a means of removing genetic material from one cell and inserting it into another. Genes are much too tiny to be manipulated directly. It would be impossible to go into a cell, snip a gene out of a chromosome, and place it inside another cell. Enzymes serve as a kind of chemical "scissors," enabling reseachers to cut and join pieces of genetic material. A variety of enzymes are available to do these jobs, as you will learn in the following sections of this book.

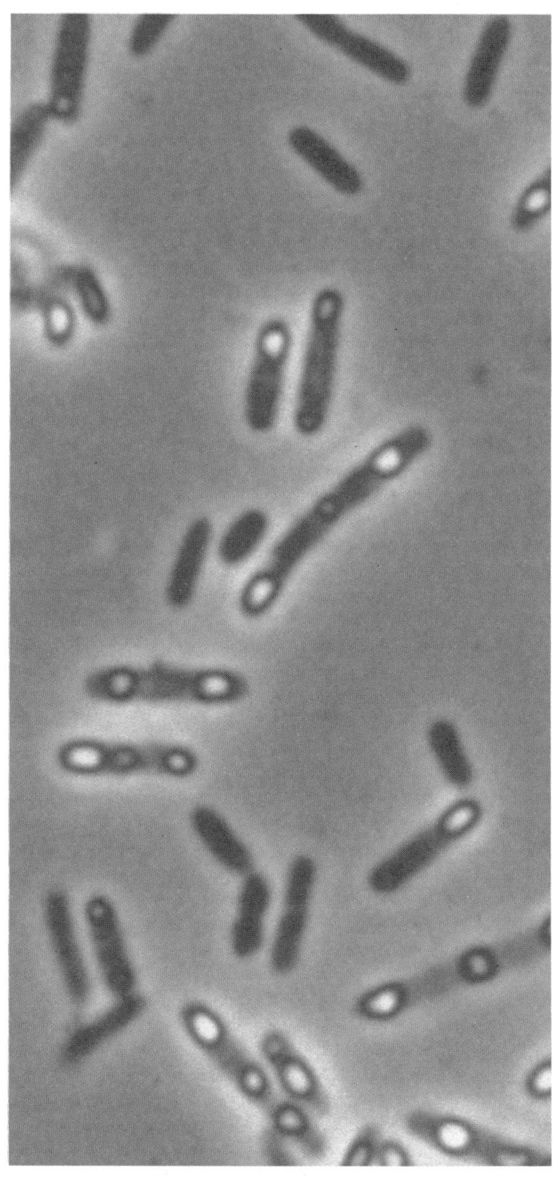

E. coli bacteria play an important role as hosts in rDNA work. These *E. coli* cells have been genetically engineered to produce a protein called interleukin-2.

PUTTING MICROBES TO WORK

Tricking bacteria and yeasts into making large amounts of proteins for which they have no use or need involves four basic steps. These are (1) obtaining the gene that codes for the desired product; (2) inserting the gene into a host microorganism; (3) forcing that microorganism to manufacture the product using the instructions on the inserted gene; (4) collecting, purifying, and testing the product.

Finding the Gene The first and often the biggest step is finding the gene that will direct the production of the desired protein. Where one looks, of course, depends on what the protein is. If it is one normally synthesized by bacteria, the place to search is the genetic material of the bacteria that make it. For example, the protein may be made by *E. coli*. The DNA of this bacteria consists of 3.4 million base pairs, forming at least 1,000 genes.

But suppose that the protein is of human origin. Then it will be necessary to find the gene among the number of functional human genes in a body cell, estimated to be between 30,000 and 100,000 and including a total of perhaps 3 billion base pairs. (This total includes large numbers of bases that do not code for proteins and whose function is unknown.)

When little knowledge is available about the location and base sequence of a gene, the "shotgun" technique may be used. This involves isolating DNA from cells of the donor organism, cutting it into pieces, and inserting the pieces into bacterial plasmids, which carry them into host cells. Then the cells containing the desired gene are identified by time-consuming, trial-and-error procedures.

Of course, researchers prefer to use more precise methods of locating the desired gene whenever possible. If the protein to be made is one that is synthesized naturally in large quantities by a particular type of cell, that cell is the place to find the genetic material needed. For the production of insulin, for example, one would look in cells from certain tissues in the pancreas.

These specialized cells contain a great deal of the messenger RNA that directs the synthesis of the protein. To obtain the necessary genetic material, the cells are broken and the mRNA extracted. Then the mRNA is used as a template to make a section of DNA exactly like that from which the mRNA was originally formed. This **copy** or **complementary DNA** (cDNA) is produced with the help of a viral enzyme, reverse transcriptase. Copy DNA is equivalent to the gene coding for the protein but without any of the extra, noncoding material that might accompany it. With some additional processing, cDNA can be spliced into bacteria and used to produce the desired protein.

If cells that produce the protein are not available, it is possible to construct cDNA with a "gene machine." The use

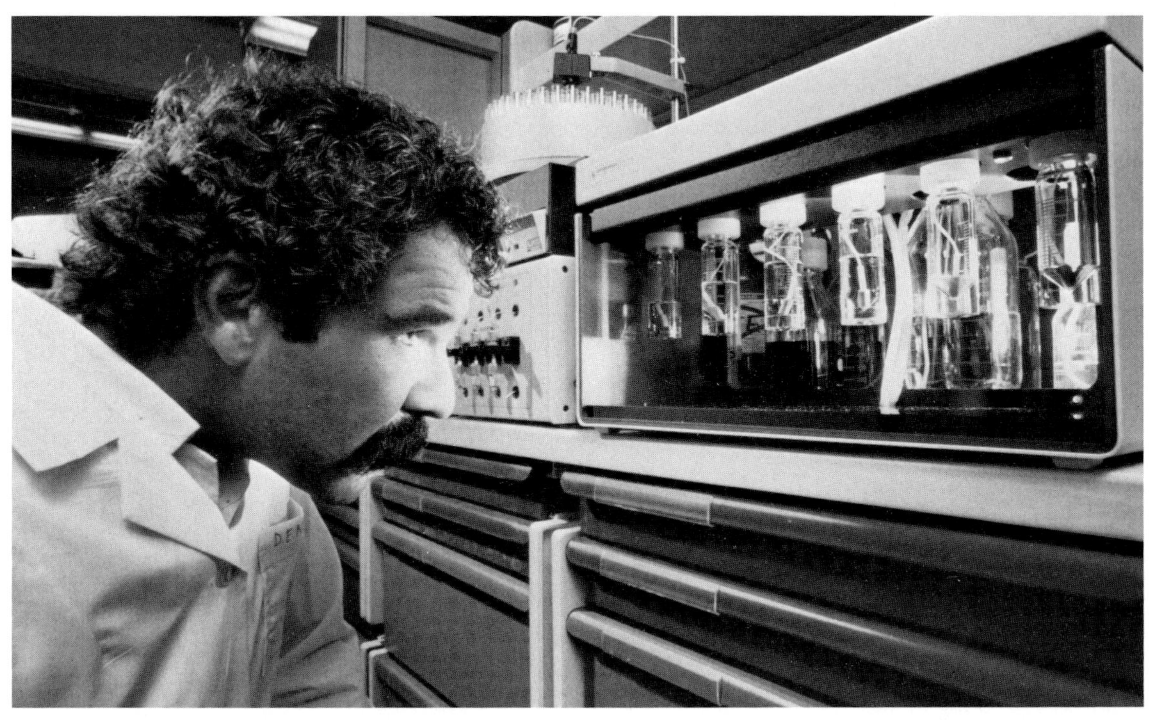

This scientist is checking the operation of a DNA synthesizer, or gene machine, an instrument that can put together specific sequences of DNA.

of this technique depends on knowing the base sequences of the DNA. Those sequences may often be obtained from computerized libraries that maintain records of genes and fragments of DNA whose base sequences have been worked out. On the other hand, if the amino acid sequence of the protein to be made is known, the genetic code can be used to figure out the base sequence of the DNA.

Once this information is obtained, the gene machine or **DNA synthesizer** takes over the job of making the cDNA. This instrument can make **polynucleotides**, which are chains of nucleotides (bases along with their associated sugars and phosphates) to exact specifications. A gene machine consists of a column of solid particles through which flow the nucleotides and necessary chemicals. Directed by computer, nucleotides can be added to the chain every 30 to 45 minutes so that the entire process takes place in hours instead of the months of work required for manual methods.

In genetic engineering, cDNA frequently

provides the gene that will be spliced into an organism to direct the production of the desired protein. Copy DNA can also be used as a "probe" to locate a specific sequence of DNA in a cell. The cDNA must contain at least some length of the base sequences that are the same as those of the gene to be retrieved. When mixed with the donor DNA, the cDNA binds to the complementary sequence. It is tagged with a tracer of some kind, often a radioactive one, that reveals the location of the sequence being probed for. Then this sequence can be removed from the rest of the DNA and spliced into another organism. (DNA probes are also used in medical diagnosis to locate defective genes that cause some hereditary diseases.)

After a gene has been located, the next big job is to separate it from the rest of the DNA. This is the work of enzymes, those remarkable biological "scissors" mentioned earlier. The enzymes used to cut a gene from the surrounding DNA belong to a class known as **restriction endonucleases**. They occur in many bacteria, where they destroy the genetic material of invading viruses.

More than 300 different restriction enzymes are known. Each one recognizes a specific sequence of bases in a double strand of DNA. For example, the restriction endonuclease *Eco* RI, which is synthesized by a particular strain of *E. coli*, cleaves the sequence CTTAAG between the bases adenine and guanine. It cuts the complementary sequence, GAATTC, between the guanine and adenine. The enzyme breaks the bonds of both strands at two points, freeing the length of DNA between that contains the gene.

DNA FRAGMENT

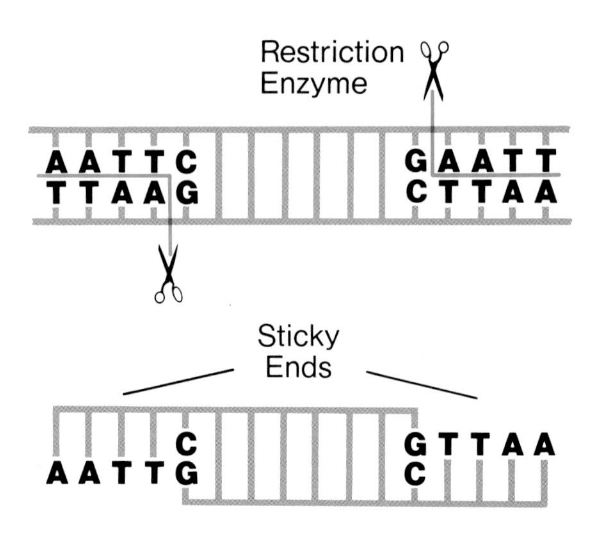

This cleaving process leaves "sticky ends," short single strands of four bases, at each end of the DNA fragment. These unpaired bases will bind to the complementary bases on another piece of DNA with sticky ends. In the case of *Eco* RI, the unpaired bases at one end of the gene to be inserted are AATT, which will bind to the complementary bases TTAA.

When the gene to be used for splicing is made up of cDNA, special "linker" bases are often added at the ends of the DNA fragment during processing. These base pairs are then cut by a restriction enzyme

PLASMID

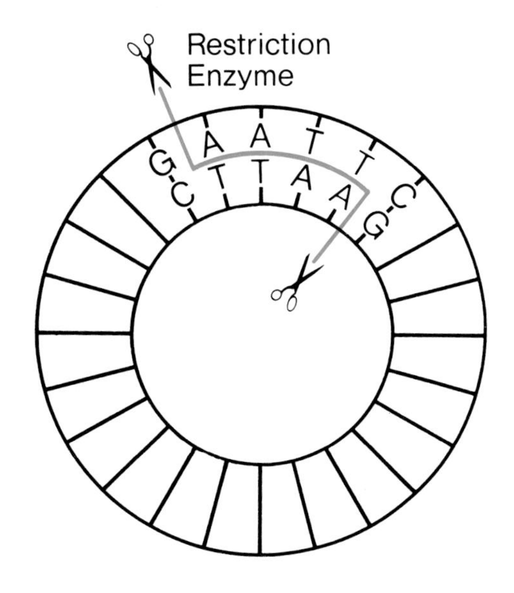

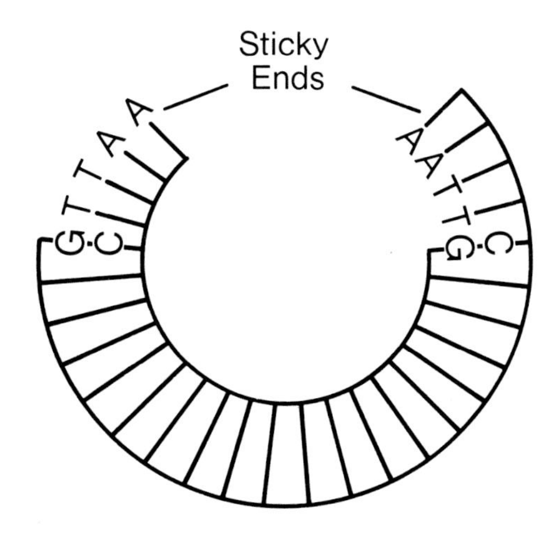

in order to create the sticky ends necessary for bonding.

A bacterial plasmid is usually the **vector** (carrier) used to receive the foreign gene and insert it into the host cell. The same restriction enzyme that cut the gene free is used to open the plasmid ring at the same base sequences in the DNA. This cutting leaves sticky ends in the opened ring that correspond to those on the DNA.

The sticky ends are bound together with the help of another enzyme, DNA ligase, which is a kind of biological glue. When the attachment is made between the ends of a fragment of DNA and those of the opened plasmid, the two pieces of genetic material are recombined. The plasmid now contains not only its own DNA but also the foreign gene that has been spliced into it—recombinant DNA.

RECOMBINANT DNA

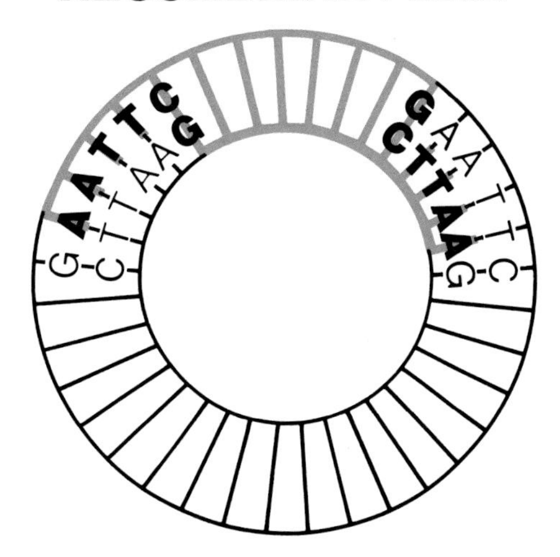

TRANSFORMATION—INSERTING RECOMBINANT DNA INTO HOST CELLS

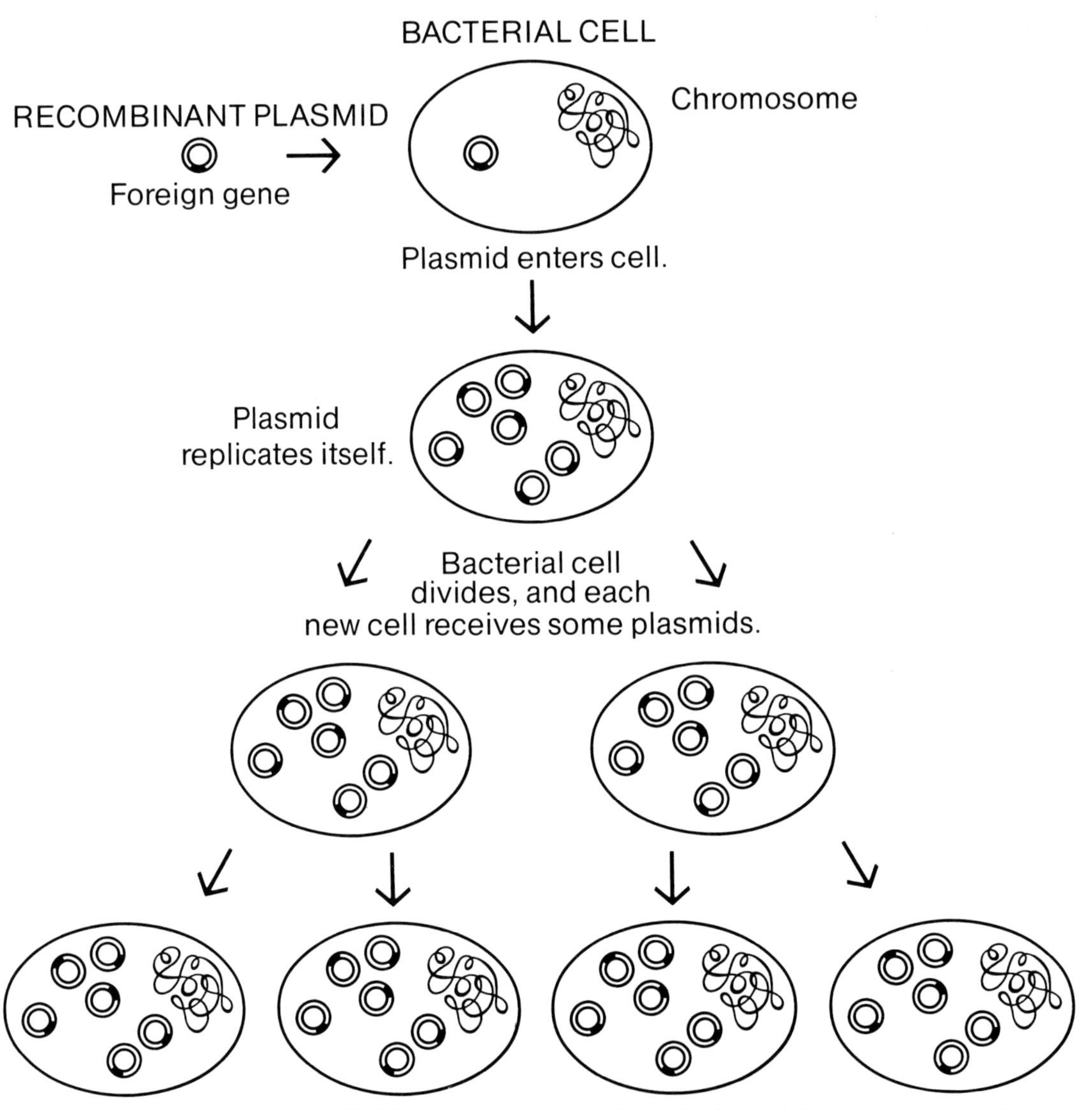

As cells continue to divide, many recombinant plasmids are produced. Each contains a copy of the foreign gene.

Inserting the Gene into a Host Cell The second basic step in the biotechnology system is inserting the recombinant DNA into host cells, a process known as **transformation**. The prepared plasmids are added to a culture of bacterial or yeast cells whose walls have been treated so that the plasmids may pass through. Once inside the host cells, the plasmids begin to replicate by much the same process as chromosomal DNA. Each time a new plasmid is formed, so is a copy of the foreign gene. When the host cells divide, each new cell receives some of the plasmids. Eventually there may be millions of the host cells, many containing plasmids with copies of the foreign DNA.

Next, the cells containing plasmids with the donor DNA must be identified. As mentioned earlier, many plasmids carry genes that make a cell resistant to certain antibiotics. This characteristic is used to identify the plasmid-carrying cells. All the cells are spread onto the surface of a solid growth medium containing the antibiotic to which the plasmid confers resistance. Cells with copies of the plasmid survive, while the ones without are killed or weakened by the antibiotic.

The plasmid-bearing cells divide until there are so many that their growth is visible. Each mass of cells, all the result of the division of a single cell and thus identical, appears as a small, raised mound called a colony. A small amount of material is taken from a chosen colony and put into a liquid medium. This will be the beginning of the next stage in the process, putting the organism to work making the product.

Putting the Organisms to Work In this stage, the challenge is to find the ideal conditions under which the cells will thrive. As they grow, they will follow the instructions on the foreign gene and produce the largest quantity possible of the desired product, along with the proteins they need for their own life processes. After the problems have been solved with small volumes of the growth medium, larger and larger batches are grown. At each stage, the problems become more complex. The ultimate goal is to devise a way of making the product on an industrial scale, which might involve a vessel that holds thousands of gallons of culture medium.

It is the job of the bioprocess engineer to provide the ideal environment for maximum yield of the product. To keep the microorganisms alive, dividing, and busily making the desired protein, the engineer must supply them with the proper nutrients, the balance of which may determine the rate of production. The culture medium must also be kept at the proper temperature and oxygen concentration and mixed to prevent the buildup of waste products. Some bioreactors being developed now use computers to monitor and adjust temperature and other conditions in the culture medium.

Another important job of the bioprocess

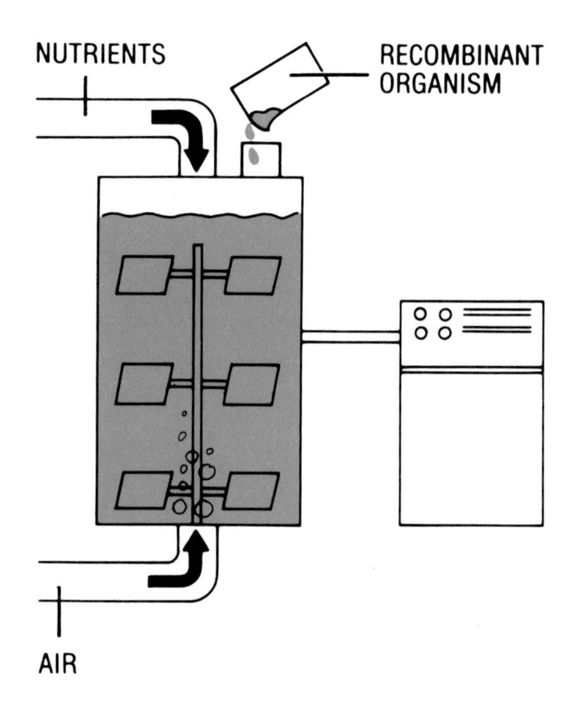

In a bioprocess, the recombinant organisms are grown in a vessel containing a culture medium. Nutrients and oxygen are supplied to keep the cells alive and producing the desired proteins.

engineer is keeping the culture free of microorganisms other than the genetically engineered ones. If contaminating bacteria or yeasts are allowed to get into the system and multiply, they can destroy the entire process, using up nutrients and possibly destroying or altering the protein being produced. Foreign organisms could even kill off the population of altered cells, which have a less than fighting chance to compete with the outsiders because of the extra load of DNA they carry.

Collecting the Product If problems can be avoided and proper conditions maintained, the culture medium will eventually contain billions or even trillions of cells, along with a large amount of the product. In the batch system of bioprocessing, the vessel is emptied, the product collected, and the tiny, hard-working "bugs" discarded. The process is then started fresh with new organisms and nutrients. (In another bioprocess known as the continuous flow system, the organisms are not discarded but are used over and over.)

The way in which the product is harvested depends both on its nature and the type of organism that made it. In the case of *E. coli*, the protein stays inside the cell. In order to recover it, the cell wall can be made leaky by chemical treatment so that the product can pass through. If this is not possible, the cells must be broken open to free the product. Other kinds of microorganisms release the protein naturally by secreting (passing) it through the cell's outer membrane and wall.

Once the protein is free of the cell, it must be separated from the medium that contains it. This is accomplished by such means as distillation (condensing vapor formed by boiling the liquid) or centrifugation (spinning at high speeds). Next, the product must be purified. Filtration and other methods are used to eliminate impurities produced by the host organisms, such as toxins from the cell walls

After the bioprocess is complete, the proteins must be separated from the culture medium. Centrifuges like the ones shown in the photo on the left are sometimes used to spin the medium at high speeds, causing the proteins and other products of the cells to collect in the bottom of the vessels.

of some bacteria.

After purification, the product is carefully tested to make sure that it is safe and effective. In the case of drugs produced by a bioprocess, this testing must be done according to procedures set by the Food and Drug Administration (FDA). The first tests are conducted on laboratory animals, followed by clinical testing on humans. Only after extensive study is a product approved for general use.

Chromatography (below) is one of the methods used to purify the products of a bioprocess. The mixture containing the protein is passed through a column filled with a material that separates its different elements into layers. As each layer moves through the material, it is collected at the bottom. By this means, the desired protein can be separated from toxins and other impurities produced by the recombinant organisms.

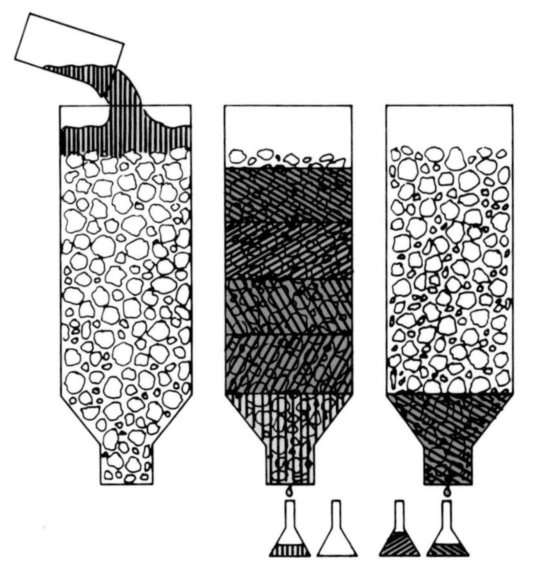

THE MICROBIAL WORK FORCE

While *E. coli* is used frequently in rDNA work, other kinds of bacteria are better suited for certain applications. *Bacillus subtilis* is a bacterial species originally isolated from soil. Because it is in no way associated with the human body and produces no toxins of the types found in varieties of *E. coli*, some researchers feel that it is safer to use. *B. subtilis* secretes proteins into the culture medium, simplifying product recovery.

Other bacteria offer different advantages. The group known as anaerobes grow in the absence of oxygen, so their medium does not have to be aerated. Thermophilic, or heat-loving, bacteria are less prone to contamination because most foreign microorganisms cannot live at the extremely high temperatures in which they thrive. In contrast, *E. coli* is happiest at human body temperature, 98.6 degrees F (37 degrees C), which is also ideal for the growth of many other kinds of bacteria.

In addition to bacteria, yeasts are being used with increasing frequency for genetic engineering. Although it is more difficult to get plasmids through a yeast cell wall and then into the nucleus, yeasts have several characteristics that make them superior to bacteria for some uses. They are capable of carrying out certain steps in the synthesis of some kinds of proteins that bacteria cannot do, for example, the addition of sugar molecules where required. Yeasts secrete their products into the medium, which saves the step of breaking open the cells. Also, yeast cell walls contain no toxins that would have to be removed from the product.

Another advantage of yeasts is that they are generally held in higher regard by the public than are bacteria. Yeasts are associated with the baking of bread and with brewing, as opposed to the narrow view many people have of bacteria as the cause of disease and spoilage. Products made with yeasts are thus likely to be met with less resistance from consumers than bacterial products.

Cells from human tissue and that of other mammals may be grown in culture to produce certain proteins, but it is a much more involved procedure than using microorganisms. Animal cells require a complex and very carefully balanced growth medium that contains a considerable number of nutrients and vitamins as well as expensive blood serum. Some need a solid surface on which to attach. The cells are fragile and easily harmed by the buildup of their own waste products. They are also prone to contamination by bacteria or yeasts.

Another disadvantage of animal cells is that they grow very slowly. In the 24 hours that it takes a culture of *E. coli* to multiply into the billions and to produce pounds of product, animal cells divide only once. Thus it takes weeks instead of days to obtain a good yield. Nevertheless, animal cells are useful for certain processes, and work is now being done

Yeast cells are also used as hosts in genetic engineering. The lab technician shown here is taking a sample of yeast cells that are producing the human protein interferon.

to find more efficient ways to grow them.

Plant cells may be cultured by methods similar to those used for animal cells. The problems involved are also similar, but plant cells too have their special uses.

CELL FUSION

Cell fusion combines genetic material from two sources, but it uses a different method than that employed in rDNA work. Instead of transferring DNA from one cell to another, two different cells are merged to form one hybrid cell containing all the DNA of both cells. The goal of cell fusion is to produce a hybrid that combines the good qualities of the original cells. To achieve this goal, cells of the same species or of different ones may be merged. Like recombination, fusion occurs in nature as well as in the scientist's laboratory.

Cell fusion can be accomplished by several different methods. The first method developed makes use of an inactivated virus that is added to a suspension of the cells to be fused. When the virus comes in contact with two cells that are touching, it causes the cell membranes to open, allowing the contents to flow together. Chemicals, particularly polyethylene glycol, are also used to partially dissolve membranes and allow cells to merge.

Fused cells are often used in research to study the location of specific genes on chromosomes. The most common application of cell fusion technology, however, is in the production of monoclonal antibodies. These products of hybrid cells have already found wide application in the diagnosis of diseases in both humans and animals.

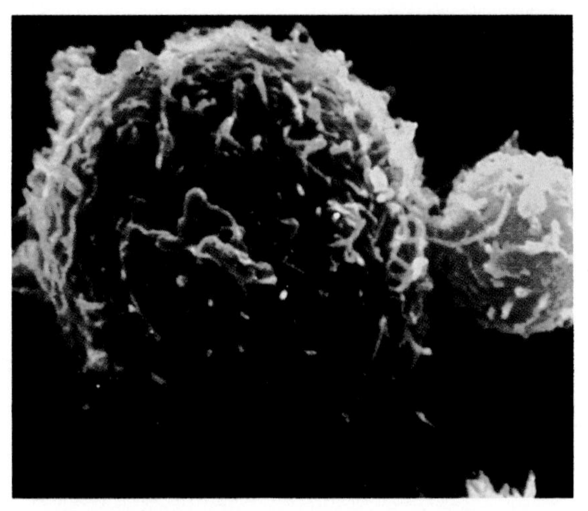

This scanning electron micrograph shows fusion taking place between a mouse myeloma cell (left) and a mouse spleen cell (right). The resulting hybrid cell, or hybridoma, can be used to produce monoclonal antibodies.

Antibodies are protein molecules produced by the immune system, the network of organs and cells that protects humans and other animals against disease. They are made in response to a foreign substance, or **antigen**, that might pose some threat to the body. When an antibody encounters an antigen such as a bacterial cell or a pollen grain, it binds to it until cells of the immune system arrive and destroy it.

An antibody molecule is extremely specific, attaching only to the antigen against which it was made. This feature is useful in diagnosing various diseases or the presence of certain drugs and other abnormal substances. When an antibody specific for a substance is mixed with that substance, an observable reaction,

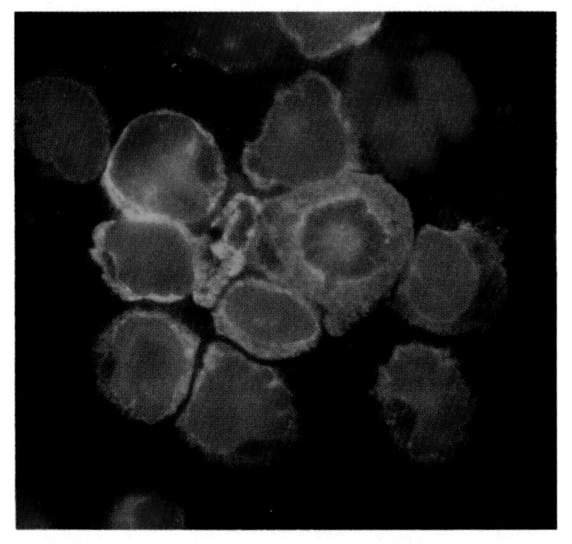

Hybridomas making monoclonal antibodies against interferon. The colors are produced by special staining methods used to show details.

such as the formation of clumps of cells, takes place.

The traditional way to produce antibodies is to inject a laboratory animal with an antigen, allow its immune system to make antibodies against the substance, and then collect the antiserum (blood serum containing antibodies). This method, however, has some drawbacks. It provides limited amounts of the desired antibody. Moreover, the antiserum frequently contains antibodies to other substances to which the animal has been exposed, which can cause false reactions in tests. Finally, batches of antiserum from different animals or collected at different times vary in composition.

Monoclonal antibodies overcome these problems. Obtained from clones of a single cell exposed to a single antigen, all the molecules are identical and recognize that antigen only. These remarkable antibodies are produced by hybrid cells created through cell fusion.

The hybrids were developed to take advantage of the best qualities of two types of cells. The first kind are mouse spleen cells that can produce antibodies but die out quickly in culture. The second kind are myeloma cells from a kind of tumor that occurs in the bone marrow of mice. These cells grow well in a laboratory and, in fact, are sometimes referred to as "immortal cells" because of their ability to reproduce themselves.

The first step in making monoclonal antibodies is to inject mice with the pure antigen for which the antibody is desired. Several weeks later, after the mouse's immune system has had time to make antibodies, spleen cells are removed. These cells are fused with cells from a myeloma by partially dissolving the cell membranes of each. The fusion results in hybrid cells, known as **hybridomas**. Hybridomas can be grown in a culture medium or in the abdominal cavities of mice. As the cells multiply, they make monoclonal antibodies.

Hybridomas made by fusing cells other than those from mice are also being developed. Human spleen cells and human myeloma cells, for example, could produce monoclonal antibodies more suitable for some applications than those made from mice cells. These kinds of antibodies are just now becoming available.

The technique of making monoclonal antibodies by means of hybridomas was developed in 1975 by two scientists, Cesar Milstein and Georges Kohler. Eight years later, the first product of genetic engineering to reach the public was a monoclonal antibody for diagnosing a disease in calves. Today monoclonals are being used in several different fields, as you will soon discover, and their future potential in biotechnology seems almost unlimited.

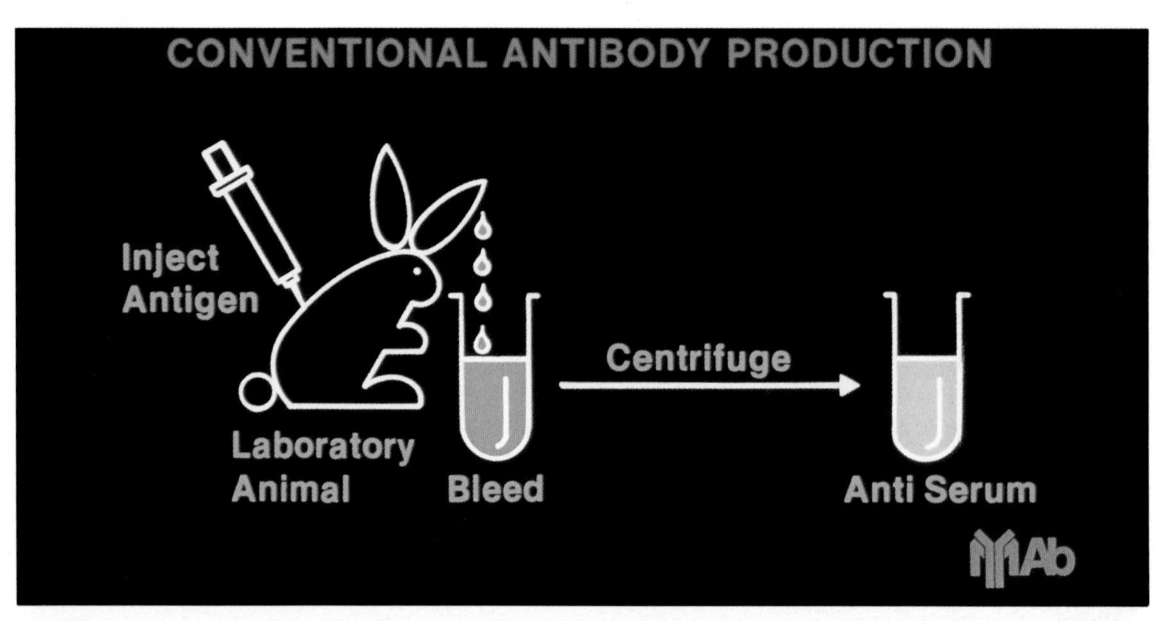
CONVENTIONAL ANTIBODY PRODUCTION
Inject
Antigen
Laboratory
Animal
Bleed
Centrifuge
Anti Serum
MAb

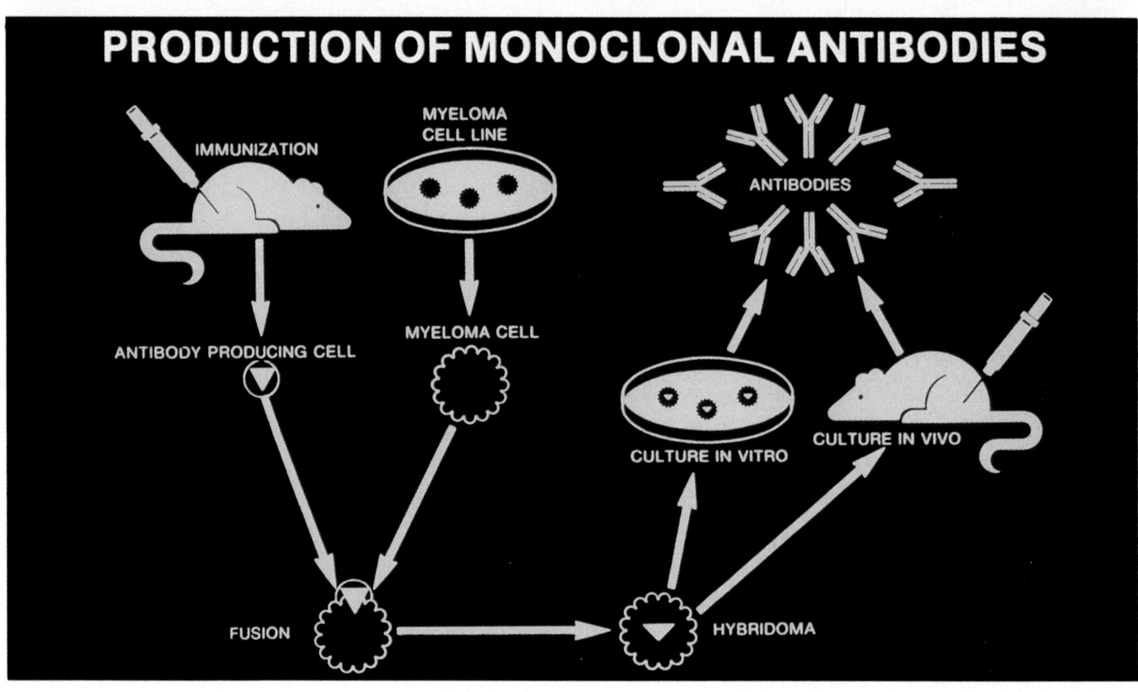
PRODUCTION OF MONOCLONAL ANTIBODIES
IMMUNIZATION
MYELOMA
CELL LINE
ANTIBODIES
ANTIBODY PRODUCING CELL
MYELOMA CELL
CULTURE IN VITRO
CULTURE IN VIVO
FUSION
HYBRIDOMA

2

A NEW AND CONTROVERSIAL INDUSTRY

The rapid development of biotechnology that has taken place in the past few years has created a whole new industry and has affected people in a variety of different professions. Thousands of scientists and technicians are now conducting research in laboratories of universities and private firms around the world. In the boardrooms of large and small companies, executives make plans for the development of new biotech products and explore ways to increase profits. Stocks of biotechnology companies are traded actively on Wall Street, while investors search for the one stock that will bring them the highest return on their money.

Various government departments and officials have also become involved in biotechnology. The U. S. Patent and Trademark Office makes decisions about granting patents to products and processes developed by biotechnology companies. Members of Congress consider legislation to fund and regulate research. Officials in federal agencies like the National Institutes of Health, the National Science Foundation, and the Food and Drug Administration review grant applications and enforce rules governing testing of products. On the state level, special commissions encourage, fund, and regulate biotechnology research.

The new biotechnology has also attracted the attention of members of the general public. People concerned over possible dangers of genetic engineering have sought court orders to halt experiments. In city and town council chambers, citizens are speaking their minds about biotechnology research carried on in their communities.

THE RESEARCHERS

The people most directly involved in the new biotechnology are the scientists who do the basic research. Their work and the way in which it is done have been directly affected by the growth of the new field.

Imagine an assistant professor of molecular biology (one of the most common specialities in biotechnology) in her laboratory at a major university. Using results from the previous week's research, she has just come up with a new theory explaining how a particular gene is regulated. While the lab technicians prepare to carry out an experiment testing the theory, the biologist scans the table of contents of a recent scientific journal. She hopes that researchers at other institutions have not won the race to produce and publish the same data she has been pursuing.

Besides the prestige she would win, being first with the answer might mean more and larger grants with which to hire more technicians and buy new equipment. It might ensure that she would be given tenure (a permanent position) at her university and possibly lead to a promotion to full professor.

The biologist quickly turns to an article in the journal that deals with a project similar to her own. After reading it, she relaxes, realizing that these researchers are moving in an entirely different direction. Returning to the lab, she checks to see how the technicians are progressing with the experiment.

Now picture another molecular biologist returning to his lab in the research complex of a biotechnology company. He has just reported to company executives the preliminary results on the production of a protein by genetically engineered bacteria. All hopes are pinned on this protein being the first product to be marketed by the firm. Its development has taken longer than expected, and the price of the company's stock is beginning to fall.

The researcher feels that he is close to a breakthrough that could save the company, not to mention earning bonuses for himself and his team and advancing his own career. He calls the lab technicians together for a pep talk. They *have* to solve the remaining problems and get those bacteria to produce.

These two scenes illustrate some of the differences that have traditionally existed between academic research and research conducted in the corporate world. Scientists at universities have tended to pursue research that is interesting to them, following lines of inquiry wherever they may lead. This freedom to explore frequently yields unexpected results—from the discovery of chemical substances to finding bacteria that get their nourishment in odd ways to learning about the actions of enzymes. Researchers working for private firms are more likely to be working toward a specific goal, trying to develop a product or process that will be profitable for their company.

In the new world of biotechnology, however, the lines between academic and corporate research are becoming blurred. A number of companies are paying for research conducted by universities. Often academic researchers are lured away from universities to work for private firms. Some have even left to start their own biotech companies. Many prominent

The growth of the biotechnology industry has created many new jobs for scientists. Shown here are two researchers at Genex Corporation analyzing DNA probes prepared in a DNA synthesizer (right).

scientists serve on the boards of such companies, giving the organizations the benefit of their knowledge while they pursue their own research.

The increasing closeness in the relationship between academic research and the corporate world has both advantages and disadvantages. It gives companies access to the latest findings in the field and may serve to bring the fruits of research—a new treatment for a disease or a vaccine for livestock—to the public more quickly. It may also mean that more money will be available for conducting further research.

Some observers fear, however, that the association between academic researchers and corporations might result in conflicts over who profits from discoveries. It might also lead to more academic research designed to make a profitable product, reducing the number of unforeseen discoveries of the kind that expand the frontiers of knowledge.

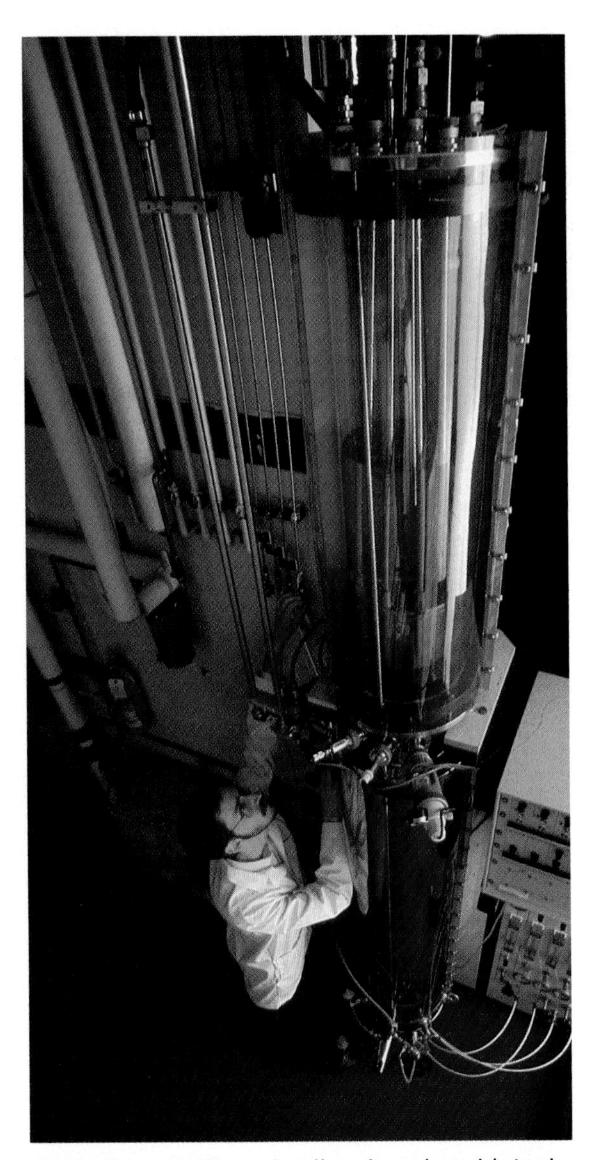

Cetus Corporation, a leading American biotechnology company, specializes in the development and manufacture of human therapeutic products. Here, a Cetus scientist is adjusting a tower fermenter in which mammalian cells are growing.

THE COMPANIES

Since techniques for altering DNA were first devised in the 1970s, hundreds of new companies have sprung up, founded by people eager to make the most of the new biotechnology's commercial potential. Many came into being before any products of genetic engineering had been manufactured or any profits made. Today few are still in existence.

Usually small in size, most biotechnology companies have had a very hard time getting established. Failures are often the result of gambling all resources on producing one particular product instead of diversifying and planning a slow, steady growth. When getting a product on the market or developing a new process takes longer than anticipated, the company may run out of funds. Then it faces failure or takeover by a larger company.

The rise and fall of companies dealing in biotechnology has been compared with the computer boom that took place in the 1970s. As with computer manufacturers, a few biotechnology companies—for example, Genentech and Cetus—have survived and become multimillion-dollar enterprises. Hundreds of other companies, however, have passed out of existence.

In addition to new companies specializing in biotechnology, many large, existing corporations have entered the field. Eli Lilly, a pharmaceutical company, and two giant chemical manufacturers, DuPont and Monsanto, were among the first to become interested in the new technology. They began research in genetic engineering with the aim of developing new products or applying the techniques to the manufacture of existing products. More companies in these areas and others such as food production and energy have become involved in biotechnology.

The majority of U. S. companies using biotechnology for commercial purposes produce pharmaceuticals or commodities for agriculture. Smaller numbers are involved in the areas of food additives, chemical production, energy, the environment, and electronics. There are also support companies that supply materials and equipment such as chemicals, DNA fragments, automation equipment, and computer software for controlling gene machines and bioreactors.

With so much at stake, competition between biotech companies is intense. In a field where being first can mean huge profits and even survival, research results and information on how products are made are jealously guarded. Securing patents to protect a company's right to microorganisms and processes that its employees have developed has become an important issue.

The patent that has had the most influence on the new biotechnology is the one granted to Stanley Cohen and Herbert Boyer for the rDNA techniques they devised in the 1970s. Scientists engaged in academic research may use these methods without paying for the privilege. Companies, however, pay an annual fee and a small percentage on the profits made by using the processes. Proceeds from the patent go to Stanford University and the University of California at San Francisco rather than to Cohen and Boyer.

Another important patent was that granted to Ananda Chakrabarty, a scientist at General Electric who developed a strain of oil-eating bacteria by combining naturally occurring plasmids. After a long legal dispute, the U. S. Supreme Court ruled in 1980 that living, altered microorganisms such as Chakrabarty's bacteria could be covered by a patent. This decision made it possible for scientists and companies to protect their genetically engineered creations and to make others pay for the privilege of using them.

THE CONTROVERSIES

Monsters created in a laboratory escaping and attacking the citizens of a city? Terrible plagues sweeping unchecked around the world, killing the human population? When word first spread that techniques had been developed for altering genetic material, some people became concerned about such possible dangers.

Fear of monsters and worldwide plagues owed more to old science-fiction movies than they did to real dangers. People with training in the sciences, however, voiced more reasonable doubts about genetic

This cartoon, published in the Minneapolis *Star and Tribune* in April 1987, reflects the belief of some that biotechnology will lead to the creation of monsters. It also expresses concern about the patenting of genetically engineered animals.

engineering. They speculated that some hybrid of bacteria and another type of organism might be able to infect laboratory workers. If such a hybrid was accidentally released from the laboratory, it might find a niche in the ecosystem and bring about some unforeseen environmental damage, even become a health hazard.

Another cause for concern among some scientists was the fact that *E. coli*, the bacterial species used most frequently for research, is a normal inhabitant of the human intestine. These people feared that a destructive genetically engineered strain of the bacteria might find its way into humans with disastrous results.

In 1974, a group of scientists proposed

a moratorium, or halt, to recombinant DNA research until further studies could be carried out. The next year, about 150 scientists met at the Asilomar Conference Center in California to discuss the potential hazards of rDNA and to plan guidelines for conducting research.

With the recommendations of the Asilomar group as a guide, the National Institutes of Health (NIH) drew up a set of regulations to govern genetic engineering. Experiments were placed into categories according to the possible risk involved. For each category, rules for the containment of microorganisms were outlined.

The NIH containment levels ranged from P1, for experiments judged least hazardous, to P4, for those considered to have the greatest potential danger. Each higher level required stricter precautions and the use of more equipment to ensure that workers were not contaminated and that microorganisms could not escape the lab. Extremely hazardous experiments such as those involving tumor-causing DNA fragments and toxins were prohibited.

In 1976, the moratorium on rDNA research was lifted, and work was resumed under the watchful eye of the NIH. The NIH guidelines applied only to institutions doing research funded by federal grants, but private companies generally chose to follow them voluntarily. Gradually, the guidelines were relaxed as cause for concern proved to be unfounded. The easing of restriction continues today, but experiments involving toxins or the release of engineered organisms into the environment must still be approved by the NIH.

Serious problems with rDNA research did not occur partly because the means of preventing such problems already existed. Techniques and equipment for conducting research safely were in use long before the question of gene manipulation arose. Microbiologists (scientists who study microorganisms) have been working carefully with potentially dangerous bacteria, viruses, and other disease-causing organisms for many years.

When dealing with this kind of material, equipment such as ventilated hoods, glove boxes, and rooms with negative pressure (to draw air in rather than allowing it to seep out) confine organisms and prevent contact with people. Before disposal, the material is sterilized in an autoclave, which kills living matter by subjecting it to steam under pressure.

Altering DNA in the laboratory was also not completely new to microbiologists. In the past, researchers produced mutations in genetic material by such means as ultraviolet light and chemicals. These agents cause random changes, some of which could be harmful. Recombinant DNA techniques have the advantage of being able to make specific, planned changes.

There are no documented cases of illness or injury caused by a genetically

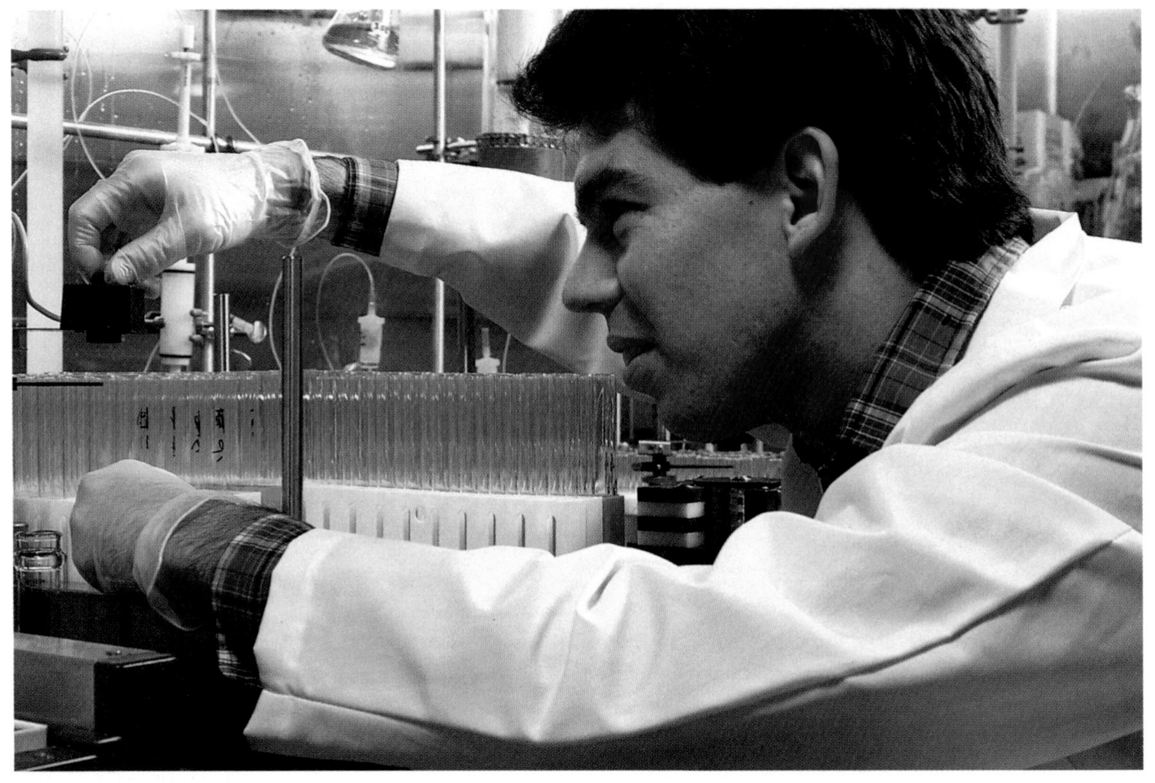

Scientists like this technician at Genentech, Inc., use the same safety precautions in working with recombinant DNA as they use in handling other potentially dangerous material in the laboratory.

engineered organism. Studies show that the microorganisms used do not survive in the bodies of laboratory workers. Most rDNA work is carried out with weakened strains of bacteria or harmless brewers yeast. Burdening organisms with foreign DNA further weakens them so that they have very little chance of surviving outside the lab if accidentally released. The extra material would make it difficult for them to compete for space and nourishment with unaltered strains in a natural environment.

There are many agencies and governmental bodies that supervise rDNA research, making sure that it is conducted safely and produces no harmful results. The Recombinant DNA Advisory Committee (RAC) of the NIH continues to rule on general matters related to biotechnology. Other agencies have jurisdiction over a specific area. The

Environmental Protection Agency (EPA), for instance, oversees the development and manufacturing of pesticides. The U.S. Department of Agriculture (USDA) has charge of animal vaccines and genetically engineered plants. The Food and Drug Administration (FDA) regulates human vaccines and pharmaceuticals. Recombinant DNA research is also controlled by state agencies and committees at universities where genetic engineering experiments are taking place.

Some people are concerned not so much with the possible hazards of genetic engineering to health and the environment as with its wider ethical implications. They fear that we have made rapid progress in rDNA technology without giving enough thought to questions of right and wrong. It troubles others that science is on the verge of understanding the mechanisms of life itself, if only its physical aspect. They see the manipulation of genetic material as "playing God" and interfering with nature.

Others do not believe that genetic recombination is in any way unnatural since it occurs frequently in nature without human help. These people argue that harnessing nature for human benefit is not "playing God" but rather a use of our God-given abilities. They also point out that humans have been creating new life forms ever since they first learned to cultivate new characteristics in plants and breed new traits in animals.

The most serious concern for some is that genetic engineering might be used to tamper with human beings, perhaps to create a super race or to produce dominant and subject groups. Of course, no one involved in the debate on biotechnology believes that this is an acceptable goal. Authorities in the field say that the ability to perform such manipulations of human genes does not exist now and may never exist. The problems to be overcome are immense. The positions and functions of only a tiny percentage of human genes are known, and scientists have very little understanding about what turns genes on and off.

The primary goal of human genetic engineering today is the treatment of a disease such as sickle cell anemia, which is caused by a single gene. The knowledge that must be gained before even this can be done is considerable. To change characteristics that are governed by many genes working together (height or weight, for example) would require technology only dreamed about today.

Still, there is good reason for caution. A great deal of research in the new biotechnology ventures into unknown territory. Any new use of genetic engineering should be preceded by a careful examination of aspects such as safety, possible effects on the environment, and ethical considerations. Many observers agree, however, that as long as rDNA research is carried out in a responsible manner, it holds considerable promise for improving the quality of our lives.

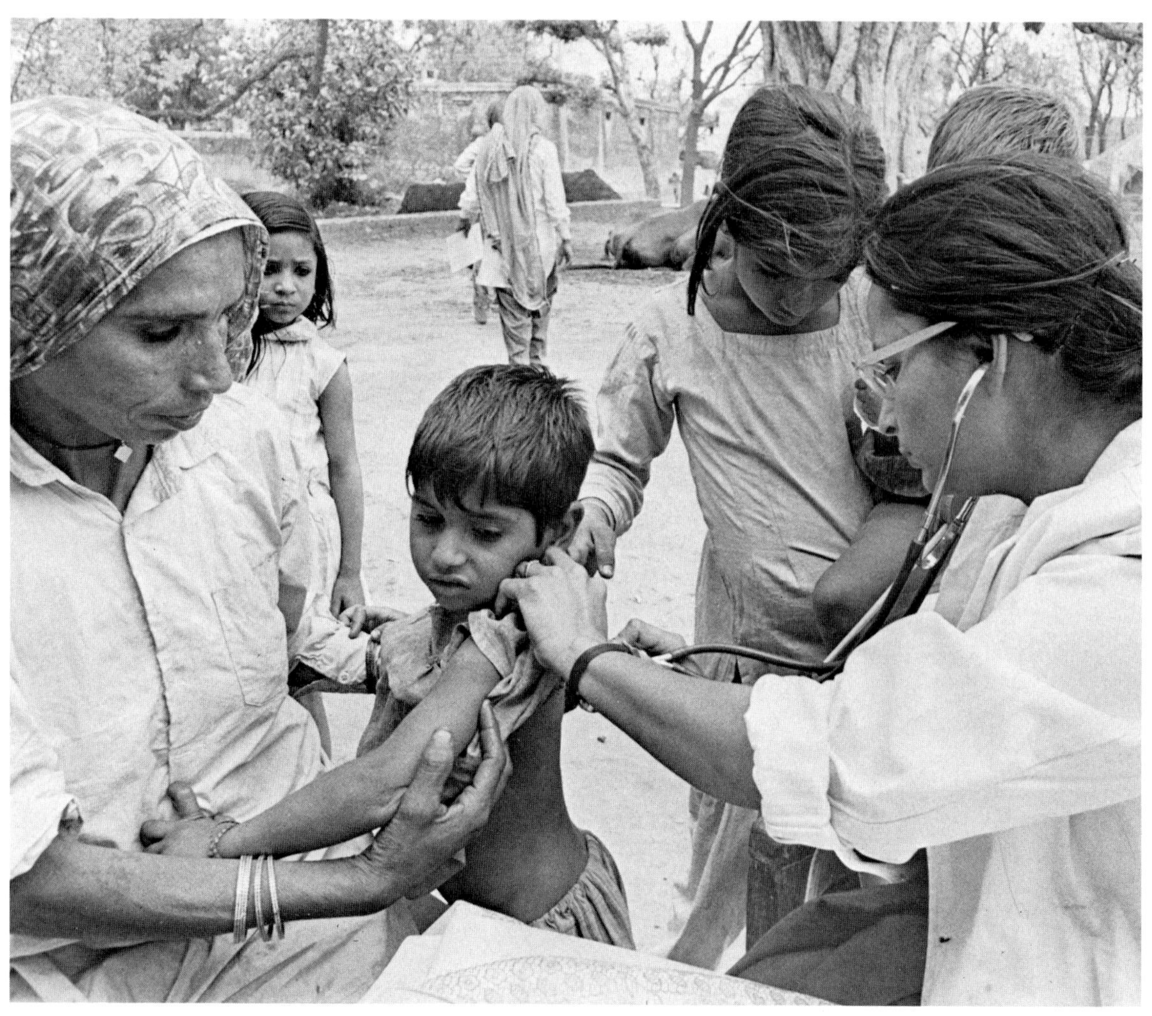

The advances in medicine made possible by biotechnology may be able to improve the health of people living in developing countries like India.

3

A REVOLUTION IN HEALING: BIOTECHNOLOGY IN MEDICINE

Not so long ago, the idea of teaching bacteria and yeasts to make human proteins would have been considered ridiculous, if not impossible. Yet today people are being treated for medical disorders with proteins produced in just this way. The pharmaceutical industry was the first to see the potential of applying the techniques of rDNA for practical uses. In recent years, companies have begun producing substances synthesized by microorganisms that supply proteins our bodies fail to make and that protect us against illness. Industry leaders hope that continuing research may eventually lead to the development of substances that will cure many disorders, perhaps even cancer, heart disease, and the common cold.

HORMONES

Hormones are proteins produced by the body's endocrine glands. They act as chemical messengers, regulating and coordinating such functions as digestion, circulation, and the concentration of minerals in the blood. Certain disorders and diseases occur when a person's endocrine glands do not make a particular hormone or fail to make it in adequate amounts. These conditions can often be treated by administering hormones that have been obtained from another source, such as animals.

It is not always easy to obtain hormones for medical treatment in large enough amounts or in a pure enough state for all who need them. Because many hormones are proteins, however, they are perfect candidates for manufacture by genetically altered bacteria and yeasts.

The first human gene product synthesized in bacteria was somatostatin, a small brain hormone consisting of 14 amino acids. It was made in the laboratories of Genentech, Inc. No medical use has yet been found for it, but its production demonstrated that rDNA methods work.

HUMAN GROWTH HORMONE

One of the first products of biotechnology to be put to practical use was human growth hormone (hGH), a protein that controls growth of bones and weight gain. Some children do not grow to a height normal for their age because their pituitary glands do not secrete enough of this protein. Sometimes these children, who might be 6 to 12 inches shorter than their classmates, begin to grow when given injections of growth hormone. The treatments must be received during their growth years, before certain bones such as the thigh bone lose their ability to elongate.

In the past, the supply of growth hormone was quite limited. The only sources for the hormone were the pituitary glands of human cadavers, carefully removed during autopsy. (Growth hormones from animals are not effective in the human body.) A week's treatment might consist of 7 milligrams (mg) of the hormone; a single pituitary gland supplies only 4 mg.

Today the supply of human growth hormone has been greatly increased because the substance is being synthesized by *E. coli* bacteria into which the gene for the hormone has been spliced. A 500-liter (110-gallon) tank of bacterial culture can make as much hGH as can be extracted from 35,000 pituitary glands. Growth hormone produced by this technique was approved for human use in 1985 and is now administered in treatment programs to as many as 5,000 children all over the United States.

Besides stimulating growth in children, there are other possible uses for human growth hormone that are being investigated. It might prove helpful in treating wounds, bone fractures, and burns as well as loss of calcium in bones of older people (osteoporosis) and some ulcers. The larger supply of the hormone available is making it possible for researchers to study these and other applications.

INSULIN

The familiar disease known as diabetes is caused by the failure of another endocrine gland, the pancreas. When this gland fails to secrete enough of the hormone insulin, the amount of sugar in the blood increases and diabetes can result. Some forms of diabetes are treated by injections of insulin. As many as 5 million people take the hormone to control their diabetes, and there are millions more worldwide who would benefit from it. The number of diagnosed diabetics also seems to be increasing.

Most of the insulin sold today comes from cow and pig pancreases collected at slaughter houses. While insulin from these animals is generally safe and effective, it differs slightly from the human hormone. A few people taking insulin from cows and pigs develop allergies to it because their immune systems recognize the animal insulin as a foreign substance. This problem is avoided by

This photograph shows the first crystals of human insulin made by recombinant DNA technology. They were produced by Eli Lilly and Company.

the use of human insulin.

Insulin was the first therapeutic rDNA product approved for sale. It went on the market in 1982. Known as Humulin,® it was developed by Genentech and is made by Eli Lilly and Company. Today, as many as half the diabetics beginning insulin treatment are given Humulin.

The biotechnology used to make human insulin is a little more complicated than the method that produces growth hormone. The insulin molecule is made up of two polypeptide chains (linked strings of amino acids), which join to make the active form of insulin. To produce genetically engineered insulin, the DNA that codes for the A chain is introduced into one batch of *E. coli* and the DNA for the B chain into a different one. The bacterial cells are induced to make the two chains, which are then collected, mixed, and chemically treated to make them link. The resulting insulin molecules are identical to those secreted by the human pancreas.

OTHER HORMONES

Research is also being done on other hormones that could be produced by genetic engineering. Among them are the neuroactive peptides, which are hormones that might be useful in pain relief and treatment of diseases of the nervous system and mental disorders. Another hormone under study is atrial natriuretic factor (ANF), which might play a role in the treatment of congestive heart failure.

PROTEINS THAT FIGHT INFECTION AND CANCER

INTERFERONS

Interferons are proteins that control the body's response to viruses and cancer cells. Belonging to a class of proteins known as immune regulators or lymphokines, they act as a kind of early warning system, alerting the immune system of an invader's presence so that healthy cells can be protected. The human body produces as many as 20 different kinds of these useful proteins.

Interest in interferons intensified in the 1950s after researchers noticed that people who caught one viral infection seldom got another at the same time. The interferons produced in response to one virus seemed to be effective in helping the body to fight off a second viral invasion. This fact made scientists think that interferons might be useful in treating a variety of viral diseases, including hepatitis, rabies, shingles and other herpes infections, influenza, and possibly even the common cold. It was also thought that they might be a weapon against certain forms of cancer, including cancer of the kidneys.

Early study of interferons was severely hampered by the lack of the material with which to work. Before genetic engineering techniques for manufacturing interferons were developed, blood and tissue cultures were the only sources. To obtain interferons from blood, the white cells are infected with a virus. But each cell produces only a tiny quantity of the protein. For instance, from 90,000 pints of blood processed in Finland during the late 1960s, only 1 gram of interferon was recovered. Its worth was estimated at $50 million. Besides being very expensive, interferons from white blood cells or tissue culture are very impure and extremely difficult to purify.

Again *E. coli* has come to the rescue. When human interferon genes are cloned in *E. coli* cells, the proteins can be made in large quantities and are quite pure. As much interferon can be produced in a liter of bacterial culture as from the white blood cells of 100 blood donors. The cost is about 10 percent of that produced by white blood cells, making it possible for researchers to obtain an adequate supply for testing. Interferon is now being produced in yeasts and in *B. subtilis* as well.

Unfortunately, research has shown that the early view of interferons as "wonder

Workers fill sterile vials with recombinant beta interferon at a Cetus plant in Emeryville, California.

drugs" was overly optimistic and that they are probably not effective against a wide range of diseases. It seems certain, however, that new ways to use the proteins against viral diseases will be discovered. Promising results have been obtained in the treatment of various types of herpes infections. There has also been some success in treating certain cancers including lymphomas and leukemias, although results against breast, lung, and colon cancer have been disappointing.

The first licensing by the Food and Drug Administration was for alpha interferon to be used as a treatment for the rare hairy-cell leukemia, which affects about 1,000 people in the United States. This decision may lead to approval of interferon to be used for other types of cancer against which clinical trials indicate it might be effective.

Just as interferon was hailed as a

KIRTLAND PUBLIC LIBRARY

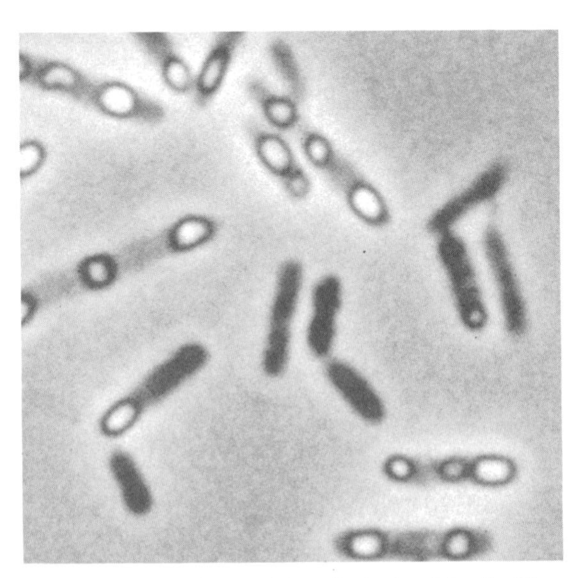

E. coli bacteria making interleukin-2. The bubble-like areas within the cells are deposits of the protein.

wonder drug when it was discovered, there are great expectations for another lymphokine, interleukin-2. First isolated from white blood cells, it has been shown to play an important role in the body's response to foreign matter. As in the case of interferon, there was not enough available for study until the gene was introduced into *E. coli* and the protein synthesized. In some early studies, interleukin-2 was mixed with the white blood cells of cancer patients, and the cells were then given back to the patients. In some cases, their tumors were reduced in size.

Other promising recombinant anticancer drugs being investigated are interleukin-1 and tumor necrosis factor (TNF).

VACCINES

As mentioned earlier, the human immune system produces proteins called antibodies that help to protect us against disease. Antibodies are made in response to foreign substances such as bacteria or virus entering the body. The use of vaccines to prevent disease is based on this **immune response**. Vaccines are suspensions of bacteria or viruses, generally ones that have been killed or weakened. When given, they challenge the immune system, which responds by making antibodies. If live viruses or bacteria of the same kind later enter the body, they are destroyed by the antibodies.

Vaccines confer active immunity, whereas passive immunity is produced by giving preformed antibodies. The antivenin serum administered to victims of poisonous snake bites contains such antibodies. Giving monoclonal antibodies also produces passive immunity, which is not as long lasting as the active protection provided by vaccines.

When making any vaccine, a major concern is that the product does not cause the disease against which it is supposed to protect. Biotechnologists are exploring several methods to make safer vaccines. Their work is based on the nature of viruses and the way in which they cause disease.

Viruses consist of a protein outer coat and an inner core of genetic material. In most viruses, it is this genetic material that is infective. The protein coat, on

One of the traditional methods of making a vaccine is growing the disease-causing organism in eggs. Producing vaccines by biotechnology avoids many of the problems involved in such methods.

the other hand, is the part of the virus that is **antigenic** (causing an immune response). Using genetically engineered bacteria or yeasts, it is possible to produce vaccines consisting only of the protein part of the virus. Such vaccines are much safer than ones made from the whole virus because they are completely free of the viral genetic material that could possibly cause the disease.

Producing vaccines by biotechnology avoids many of the problems involved in the traditional methods of production. To make a vaccine in the traditional way, the causative organism must be grown in the laboratory. Many disease agents are difficult or impossible to cultivate. Even if they can be grown, it is often hard to separate them from any substances that might be harmful.

Subunit vaccines consisting only of the parts of the organism that trigger antibody production do not have this drawback. Such vaccines may also prove to be more stable than conventional kinds. If they do not need refrigeration, they would be of considerable value in undeveloped areas that do not have electricity. These are the regions where preventive measures against diseases are most often needed.

Bacteria and yeasts are being genetically altered to manufacture vaccines against a variety of infectious diseases. A yeast-produced vaccine is being developed against hepatitis B, a serious liver disease that strikes millions every year. It is hoped that this vaccine will replace the one now available, which has many disadvantages. Made from blood products, its manufacture exposes the preparers of the vaccine to the disease. In small supply and expensive, its safety is also in doubt.

Biotechnologists are at work on vaccines against other difficult-to-treat diseases, including AIDS, chicken pox, influenza, and the common cold. A herpes vaccine is undergoing tests in laboratory animals, as well as one for Rocky Mountain spotted fever. Improved vaccines against polio and rabies are also being developed.

Another approach to making vaccines by biotechnology is to modify the **vaccinia virus**. This is the virus that has been used for almost 200 years to vaccinate against smallpox, eradicating this formerly dreaded disease. Vaccines against herpes simplex, hepatitis B, malaria, and influenza in humans, vesicular stomatitis (a cattle disease), and rabies have been constructed by genetically altering the vaccina virus. Antigenic genes from the causative agents were spliced into the DNA of the vaccinia. When a person is immunized with the vaccine, the genes enter the body cells, which make copies of the antigenic material. Alerted by the presence of the pathogens, the person's immune system then manufactures antibodies for strong, long-lasting immunity against the disease that they cause.

There are plans to develop a single vaccine against more than a dozen different diseases or various strains of one disease using the vaccinia virus. If it could be administered by jet gun, immunization of large numbers of people could be carried out by nonmedical personnel under nonsterile conditions at a cost of pennies per dose. This all-purpose vaccine could be custom tailored for the region of the world in which it would be used.

Malaria is the world's most prevalent infectious disease—there are an estimated 300 million cases a year. It is caused by various species of *Plasmodium*, which belong to the group of one-celled animals known as protozoa. *Plasmodium* protozoa have very complex life cycles, and there has been little success in growing them in the laboratory. One stage of the cycle, known as the sporozoite, involves reproduction inside the *Anopheles* mosquito, and the only source of it has been mosquito salivary glands. Because of the complexity of the disease and the difficulty of studying it, no vaccine has been developed to use against malaria. To make

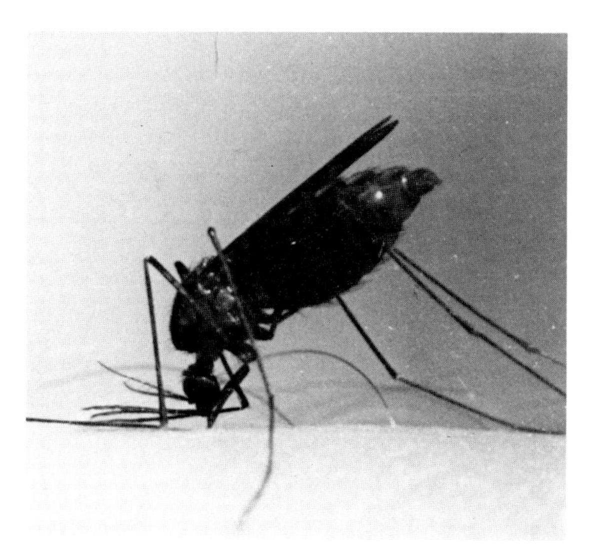

Malaria, which is spread by the *Anopheles* mosquito (above), is common in many countries of Asia and Africa (right). A vaccine developed by biotechnology could protect millions from this destructive disease.

matters worse, the mosquitoes are becoming more resistant to insecticides and the protozoa more resistant to drugs.

Various biotechnological techniques are being studied to devise a safe, effective vaccine against this destructive disease. A promising beginning was made when researchers successfully cloned genes that code for a major surface protein of the sporozoite stage in two species of *Plasmodium*. Scientists are optimistic that these proteins will prove helpful in developing a malaria vaccine.

Other diseases caused by human parasites, such as hookworm, schistosomiasis, trachoma, and amebic dysentery, affect hundreds of millions of people, especially in developing countries. There are currently no vaccines to prevent them, and treatment is difficult because many of the available drugs are very toxic. There is hope that rDNA technology will develop these much needed vaccines, but a great deal of work remains to be done.

ENZYMES

Enzymes make up another of the useful groups of proteins produced by living organisms. A variety of enzymes have been found to be valuable in the treatment of diseases. Three of these combat one of the modern world's leading causes of death, heart disease. **Streptokinase** and **urokinase** have been used for some time to dissolve blood clots in veins and arteries that block blood circulation, causing heart attacks.

Streptokinase is a natural product of *Streptomyces* bacteria. It is effective against clots, but its long-term use may cause an allergic reaction. Urokinase, a human enzyme currently obtained from urine or from tissue culture, does not have this drawback but is extremely expensive. Researchers have reported success in manufacturing it in bacteria.

Another human enzyme, **tissue plasmogen activator**, or t-PA, can be made by genetically engineered bacteria and yeasts as well as in tissue culture. It has certain advantages over streptokinase and urokinase. Because it acts only on the blood clot, it does not cause excessive bleeding as the other enzymes are prone to do. It is also effective in low doses. While streptokinase has to be administered by a catheter directly to the heart, t-PA can be injected into a vein, saving valuable time during a heart attack. It may also prove effective against strokes.

Another enzyme, alpha-1 antitrypsin (AAT), is being used to treat sufferers of emphysema, a disease of the lungs. Formerly obtainable only from human blood, it can now be manufactured by yeast cells. With a supply available for experimentation, researchers hope to find ways to use AAT to treat the effects of smoke inhalation and other respiratory problems.

BLOOD PROTEINS

The clotting of blood is a complex process involving the action of a sequence of blood proteins. Certain clotting disorders are caused by a flaw in a person's genetic material, with the result that a particular protein is made incorrectly or not made at all. One of the best known of these conditions is hemophilia, in which the individual is subject to life-threatening bleeding if injured.

There are two types of hemophilia. The more common type A is caused by the absence of the protein factor VIII; type B is the result of a deficiency of factor IX. In many cases, both these forms of the disease can be controlled by supplying the missing protein. Both proteins are currently obtained from donated blood, and Factor VIII may also be produced in cell culture. Biotechnologists are investigating ways to make them cheaply and easily by using recombinant DNA methods.

Another blood protein, human serum albumin, is used in large quantities during surgery and as a treatment for shock, burns, and protein deficiencies. While readily available from blood, it has been successfully produced in engineered bacteria and yeasts by Genentech.

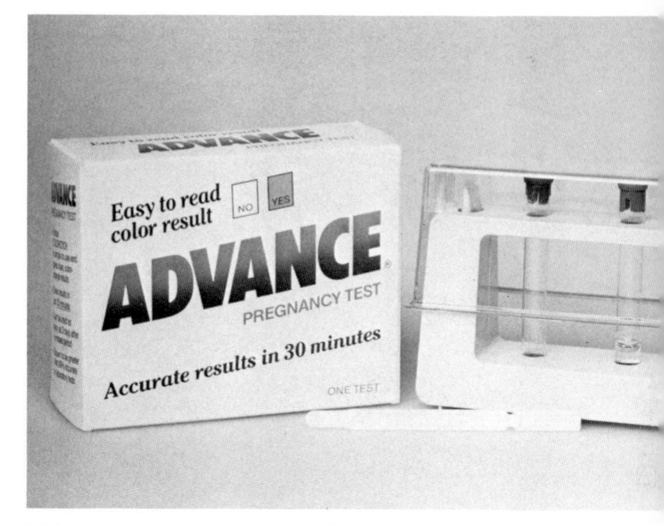

This pregnancy test kit using monoclonal antibodies is one of the few products of biotechnology available to the general public.

MEDICAL USES OF MONOCLONAL ANTIBODIES

As explained earlier, monoclonal antibodies are proteins produced by means of cell fusion rather than the techniques of recombinant DNA. Current research suggests that these products of biotechnology will play as important a role in the medical revolution as those produced by genetically altered bacteria and yeast. Monoclonal antibodies are being employed in an ever-increasing number of ways, including diagnosis and treatment of diseases, the locating of cancerous growths in the body, and purification of protein.

One of the most common ways in which monoclonals are used today is in diagnostic kits for the detection of various diseases and conditions in both humans and animals. Such kits include antibodies that react to specific substances such as a certain kind of bacteria. They are inexpensive, safe, and often much more reliable than the older methods of identifying diseases. Most diagnostic kits are now being used by doctors, but some are available to the public, including fast, very accurate tests that indicate pregnancy by reacting to the presence of certain hormones in a woman's urine.

Monoclonal antibodies are used not only to recognize diseases but also to locate tumors and other abnormal growths in the body. Certain cancerous tumors produce protein molecules not normally found in the blood. These proteins can be detected by monoclonal antibodies programmed to react to them. Such tests

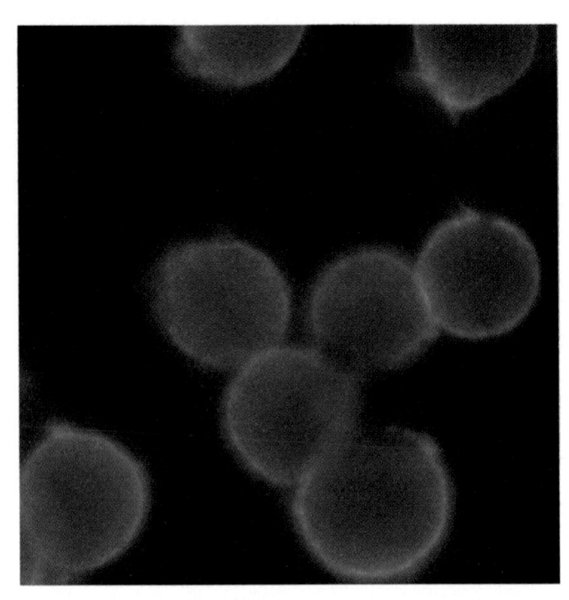

Monoclonal antibodies binding to breast cancer cells. The bright green areas show where the antibodies have attached themselves to the cells.

make possible early detection and identification of the tumor. By knowing exactly what kind of tumor is present, appropriate treatment can be started promptly. The antibodies can also be tagged with a radioactive label that will reveal the location and size of the tumor and enable doctors to detect the possible spread of cancerous cells.

In addition to cancer detection, monoclonal antibodies may have a role in cancer treatment. They might be used to deliver cancer-fighting substances directly to the affected cells. The technique would work in this way. Standard chemotherapy drugs or radioactive elements would be attached to a monoclonal antibody specific for a particular type of cancer cell. Other cytotoxic agents (those poisonous to cells) such as diphtheria toxin, cobra venom, or plant protein toxins could also be used. Carrying the agent, the antibody molecules would rush to the cancer cells and attach to them, delivering the drug directly while minimizing contact with and damage to healthy cells. This method of fighting cancer seems promising, but a great deal of testing remains to be done before it will be available for use on humans.

Another medical application of monoclonal antibodies is in helping to prevent the rejection of transplanted organs. Ortho Pharmaceutical Corporation is now manufacturing an antibody called Orthoclone OKT*3, which kills some of the T-cells produced by the immune system. T-cells attack foreign tissue in the body, including transplanted organs. OKT*3 has been shown to be effective in cases where transplant patients have suffered severe episodes of rejection of new livers or kidneys.

Monoclonal antibodies also have a useful role in purifying proteins made by other processes. Antibody molecules specific to the protein needing purification are fixed onto a solid support (glass beads, for instance) in a column. Then the fluid containing the protein is washed through and the protein binds to the molecules. After the protein has been captured, conditions are altered so that it is released from the antibodies. While a great deal

Monoclonal antibodies can fight cancer by delivering destructive substances directly to affected cells. The castor bean plant shown above produces a toxin called ricin, which could be used to destroy cancer cells.

of work remains in order to perfect this method, it is already in use to purify certain proteins, including interferon made by genetically altered organisms or in cell cultures. This is only one of the ways in which the technologies of rDNA and cell fusion could be used together to prevent disease and improve health care.

A NEW WORLD OF MEDICINE?

Biotechnologists do not promise that proteins made by these revolutionary techniques will prove to be wonder drugs that will prevent or wipe out all human diseases. Most researchers are certain, however, that proteins created by biotechnology have the potential to treat injuries as well as to cure some diseases and prevent many more at less cost and more safely than many present methods.

What is just as important is that the new technologies provide, for the first time, adequate amounts of many proteins for analysis, experimentation, and testing. It is impossible to predict what medical advances may result from such research.

4

BIOTECHNOLOGY AND THE FOOD WE EAT

Every day, people all over the United States and in other parts of the world eat a lunch made up of a cheeseburger, French fries, and a soft drink, either regular or diet. There is no item in this meal that is not now or will not soon be affected by the new biotechnology.

Of course, there is nothing new about using technology in the production and processing of food. For thousands of years, humans have worked to produce the largest possible amount of food from the resources available. Past methods of increasing production included planting seeds from the best plants to improve crops and breeding only the best animals to produce stronger, more productive livestock. Gradually, these techniques have expanded into more complex, more efficient methods applied to every aspect of agriculture.

The rate of agricultural progress has accelerated through time, reaching a peak in the Green Revolution that began during the 1960s. This ongoing revolution has brought about massive increases in yield per acre by the use of hybrid seeds, more productive strains of cereal crops, synthetic fertilizer, expanded use of chemical pesticides and herbicides, and advanced irrigation methods. Similar increases in the production of meat and milk have come about through advances in animal agriculture such as improved nutrition and artificial insemination.

Biotechnological methods for increasing the food supply will build upon the earlier methods, especially those of the Green Revolution. They may be able not only to produce more food but also to remedy some of the problems caused by certain agricultural practices. It is possible that genetic engineering will have its greatest impact in agriculture. Some observers believe that it may bring about a revolution of its own if the new techniques come into widespread use.

Such a revolution may be sorely needed in the near future. The world's population is growing, and every year, there are more mouths to feed. At the same time, the amount of land available to grow food is decreasing. In Africa, deserts are taking over formerly arable land. All over the world, irrigated crop land is suffering from a buildup of salts that prevents or slows plant growth. In developed and developing countries, the amount of farm land diminishes daily as it is taken over for building cities. In the future, techniques to grow more food on less land will be vital in order to provide nutrition for the world's billions.

PLANT AGRICULTURE

Let's take a look at the typical lunch described earlier. Making the hamburger bun involves a bioprocess, using yeast to make the dough rise. Today, biotechnology researchers are working on ways to engineer new yeast strains for improved performance.

But take the production of the bun back several steps, to growing wheat for the flour. As with all crops, the goal in growing wheat is to produce a high-quality product with the greatest possible yield per acre, while keeping costs low. In the past, growing conditions were altered to fit the requirements of the plant by such means as fertilization, irrigation, and adding lime to the soil. Today researchers are exploring methods to adapt the plant to fit its environment.

Formerly, the only way to alter plants was by crossbreeding. Individual plants with desirable characteristics were selected and bred with each other to produce new strains with the same characteristics. This method is time consuming since it requires growth to maturity to obtain the results. Biotechnology is devising ways to change the genetic material of plants directly.

Manipulating the DNA of plants is still in its early stages. Investigations have been hampered by the lack of an all-purpose vector such as the bacterial plasmid to carry genetic material between organisms. For one major type of plants, the dicotyledons or dicots (including legumes like soybeans and peas), there is a well-known vector. It is the **Ti** (for tumor-inducing) **plasmid** found in *Agrobacterium tumefaciens*, a species of bacteria that causes the crown gall tumor in some plants.

Experiments have shown that the Ti plasmid can be altered by inactivating the tumor-inducing gene and splicing in a foreign gene. When the plasmid is introduced into a plant, both the plasmid DNA and the foreign gene are integrated into the plant chromosomes. Early experiments using the Ti plasmid succeeded in inserting a bean gene into a tobacco plant and making a light-activated gene from a pea plant function in a petunia. While there were no practical uses for such altered plants, they proved that the method works. Later experiments with the Ti plasmid produced more useful results, including plants that are resistant to certain herbicides.

For many years, scientists believed that the Ti plasmid could serve as a vector

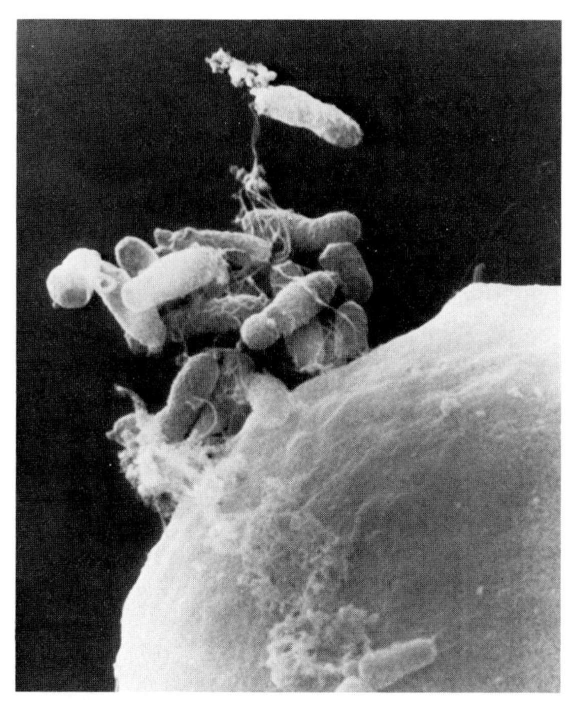

Agrobacterium tumefaciens infecting the cell of a carrot plant. When these bacteria enter a plant cell, their genetic material becomes a part of the plant's DNA.

only for plants in the dicot group. They did not think that *Agrobacterium* was capable of infecting corn, wheat, rice, and other important food crops in the monocot plant group. Then in 1987, researchers succeeded in using the bacterium to introduce a virus into corn. These results suggest that the Ti plasmid might also be used as a vector to alter other important cereal grains such as rice and wheat. Ways of synthesizing vectors are being studied as well.

In 1987, scientists discovered that *Agrobacterium* could infect corn and other monocot plants as well as dicots like carrots. This discovery opened up new possibilities for genetic engineering of crop plants.

Plant scientists are exploring other means of gene transfer in addition to the use of vectors. One experiment employs high-voltage electrical pulses to open pores in the membranes of plant cells from which the walls have been removed so that foreign DNA can pass through. This has been accomplished with carrot cells. In other research, complete chromosomes or pieces of them are injected with a fine glass needle into the cell after wall removal. This is a useful technique when several genes to be transferred are grouped on a single chromosome.

The ability to manipulate the DNA of plants could lead to significant changes in various areas of agriculture. Here are some of the possibilities being explored.

The roots of this soybean plant bear nodules formed by the nitrogen-fixing bacteria *Rhizobium.*

NITROGEN FIXATION

Wheat and many other food crops require a great deal of nitrogen to make all their protein structural components, enzymes, and genetic material. The air we breathe is 80 percent nitrogen gas, but plants cannot use nitrogen in this form. It must be fixed—combined either with hydrogen to form ammonia or with oxygen to form nitrates. Plants themselves cannot fix nitrogen; only certain kinds of bacteria and some blue-green algae (one-celled plants) have this ability.

The group of nitrogen-fixing bacteria most important to agriculture is *Rhizobium*, which has a very special relationship with legumes. *Rhizobium* cells enter the root of a legume and multiply into the millions, forming a lump known as a nodule. The bacteria then fix nitrogen by means of enzymes called nitrogenases. This is a symbiotic relationship (one advantageous to both partners) in which the bacteria supply nitrogen and the plant provides energy and shelter.

Legumes have been used as far back as Roman times to enrich the soil. They are essential to the practice of crop rotation. Alfalfa, clover, or another legume is grown in a field depleted of nitrogen by a crop such as wheat or corn, then plowed under to decay and release its fixed nitrogen. Many legume seeds are inoculated with *Rhizobium* to ensure adequate numbers of bacteria.

Another species that can fix nitrogen is *Klebsiella pneumoniae*. Unlike *Rhizobium*, it lives free in the soil and does not form nodules on plant roots.

Nitrogen-fixing bacteria are not able to infect the main monocot food crops, such as corn, wheat, rice, and animal forage grasses. These crops quickly use up the nitrogen in the soil, and fertilizers in ever-increasing amounts must be added. Chemical fertilizers are expensive. Their cost is dependent on the price of petroleum products because the hydrogen for the ammonium salts in them comes from oil or natural gas. (In the United States, 10 percent of the oil supply is used for fertilizer.) Besides the expense, their use can cause problems. Fertilizer runoff contaminates streams and lakes, promoting the growth of undesirable plants that use up nutrients and oxygen needed by aquatic life.

Several lines of research are in progress to enable monocots to fix nitrogen. The more difficult approach is to alter the genetic structure of plants so that they can fix nitrogen themselves. The **nitrogen fixation (nif) system** is quite complex. That of *Klebsiella*, the one most throroughly studied, involves 17 genes. Not only must these genes be inserted into the plant's DNA, but they must also be made to function. The *nif* genes have been successfully transferred to *E. coli* and yeasts, but it will take some time to make them function in plants.

A more promising approach to the nitrogen-fixation problem is by using the *Rhizobium* bacteria. Each species of *Rhizobium* can form a relationship only with one particular plant. It might be possible, however, to genetically alter the bacteria to enable them to fix nitrogen for plants other than the ones they normally serve. This method of altering the nitrogen-fixing bacteria rather than the plant itself will likely be the one by which the most progress is made in the next few years. It is less difficult to engineer bacteria than plants, and the procedures are already well worked out.

SALT TOLERANCE

One way to grow more wheat, or any other food crop, is to enable it to grow in a place previously unsuited for its cultivation. One such environment might be soil of high salinity (salt content). Developing plants to grow in such soil would overcome a negative effect of irrigation. While irrigation has made agriculture possible on millions of acres lacking a natural water supply, it has also ruined a great deal of land. As the water evaporates, it leaves behind the salts that it carries. In areas with some rainfall or snow melt, much of the salt is flushed away, but this does not occur in most desert areas.

One estimate is that by the end of the

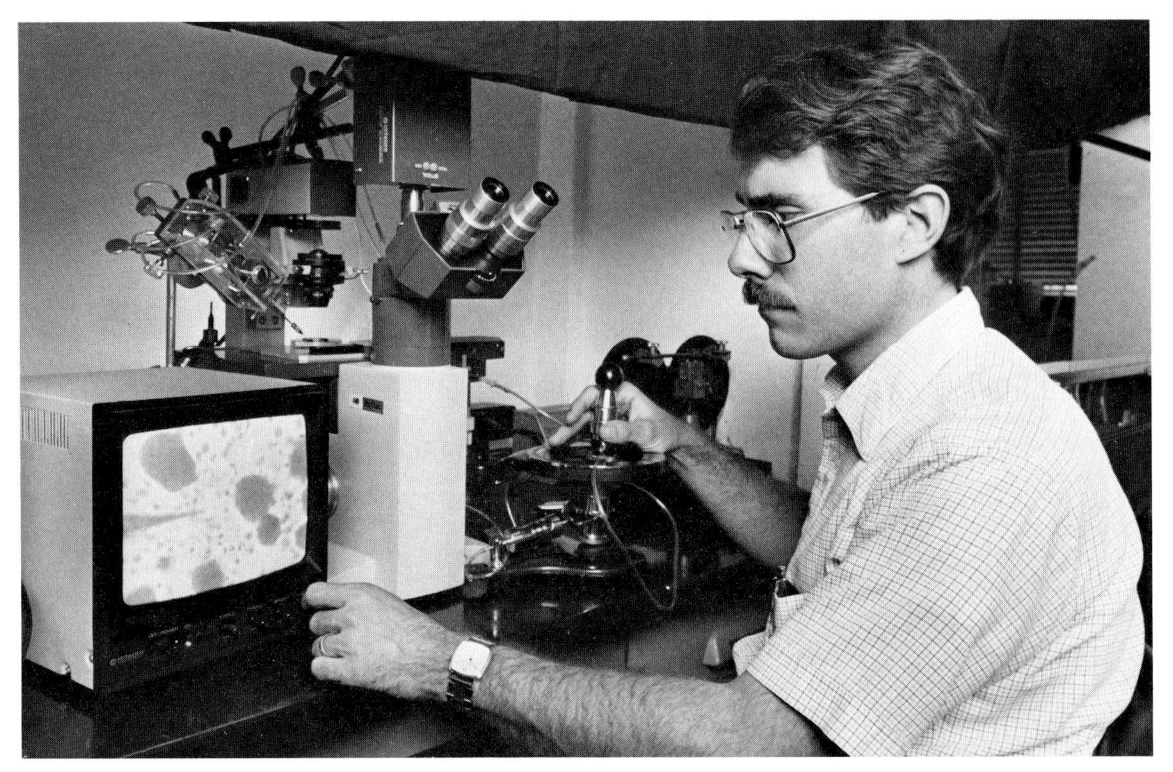

This plant scientist is using a fine glass needle (seen on monitor) to inject genes for salt tolerance into the cell of a petunia plant. If the genes successfully merge with the cell's DNA, petunia plants produced from the altered cell should be able to grow in salty soils.

20th century, 50 percent or more of currently irrigated land could be lost to cultivation because of salinity. The problem is of real concern in California and other western states dependent on irrigation for crops. In other parts of the world, there are also coastal lands made salty by periodic flooding of seawater.

Most crop plants have a low tolerance to salt. If plants could be altered to be more salt-tolerant, however, many acres could again support food production. This may be accomplished through genetic engineering. The genes could come from plants such as those that grow in salt marshes along the East Coast of the United States. Similar gene transfers might also be used to adapt plants for arid regions, areas with high concentrations of certain metals in the soil, or with other growth-limiting factors.

RESISTANCE TO HERBICIDES AND PESTICIDES

Chemical herbicides are generally unselective about which plants they kill; they go after crops and weeds alike. If a crop plant could be made nonsusceptible to these chemicals, fields could be treated with herbicide after planting to kill weeds that compete with the plants for space, nutrients, and moisture. Such resistence would also prevent problems when using fields previously sprayed.

A beginning has been made in engineering nonsusceptible plants. A mutated bacterial gene conferring resistance to the herbicide glyphosate, found in such widely used products as Roundup® or Kleenup® has been introduced into tobacco by using the Ti plasmid from *Agrobacterium tumefaciens*. (Appropriately enough, Monsanto, the chemical company that manufactures Roundup, produced the resistant plants.) Researchers are now working to make corn, tomatoes, soybeans, and other plants herbicide resistant.

Like herbicides, pesticides are essential to modern agriculture, but they can create problems. They may destroy the balance of insect populations by killing beneficial as well as harmful insects. This creates an imbalance that can result in outbreaks of insects suddenly uncontrolled by natural enemies.

Most pesticides are toxic to humans and must be used with extreme care. They are often misused, and every year, thousands of people are poisoned by them, some fatally. It is estimated that only 0.1 percent of the insecticide applied in agriculture reaches the target insect. The rest contaminates the air and soil, and runs off into streams and lakes. Some of it may remain on the product to reach the consumer.

Through the techniques of biotechnology, corn has already been given the ability to protect itself against one of its hungriest insect enemies, the root cutworm. This was done not by altering the plant but by engineering *Pseudomonas fluorescens*, a bacterium that lives, among other places, on corn roots. A gene from another species of bacteria, *Bacillus thuringensis*, was transferred to the *Pseudomonas. B. thuringensis* produces natural insect toxins and has been sold for years as a pesticide to spray on leaves. When corn seeds coated with the altered *Pseudomonas* germinate, the bacteria colonize the plant roots. The new gene they carry codes for a protein that kills the root cutworm when it begins to eat the root. Tests in laboratory and greenhouse have demonstrated that the protein is not harmful to birds, mammals, or fish.

Natural pesticides like that produced by *B. thuringensis* are very specific. The toxins are deadly to particular pests and do not kill other insects. Monsanto, the company that developed the altered *Ps. fluorescens*, plans to add additional genes to the bacteria so that they will produce toxins against other insect enemies.

PRODUCING HERBICIDE-RESISTANT PLANTS USING THE TI PLASMID

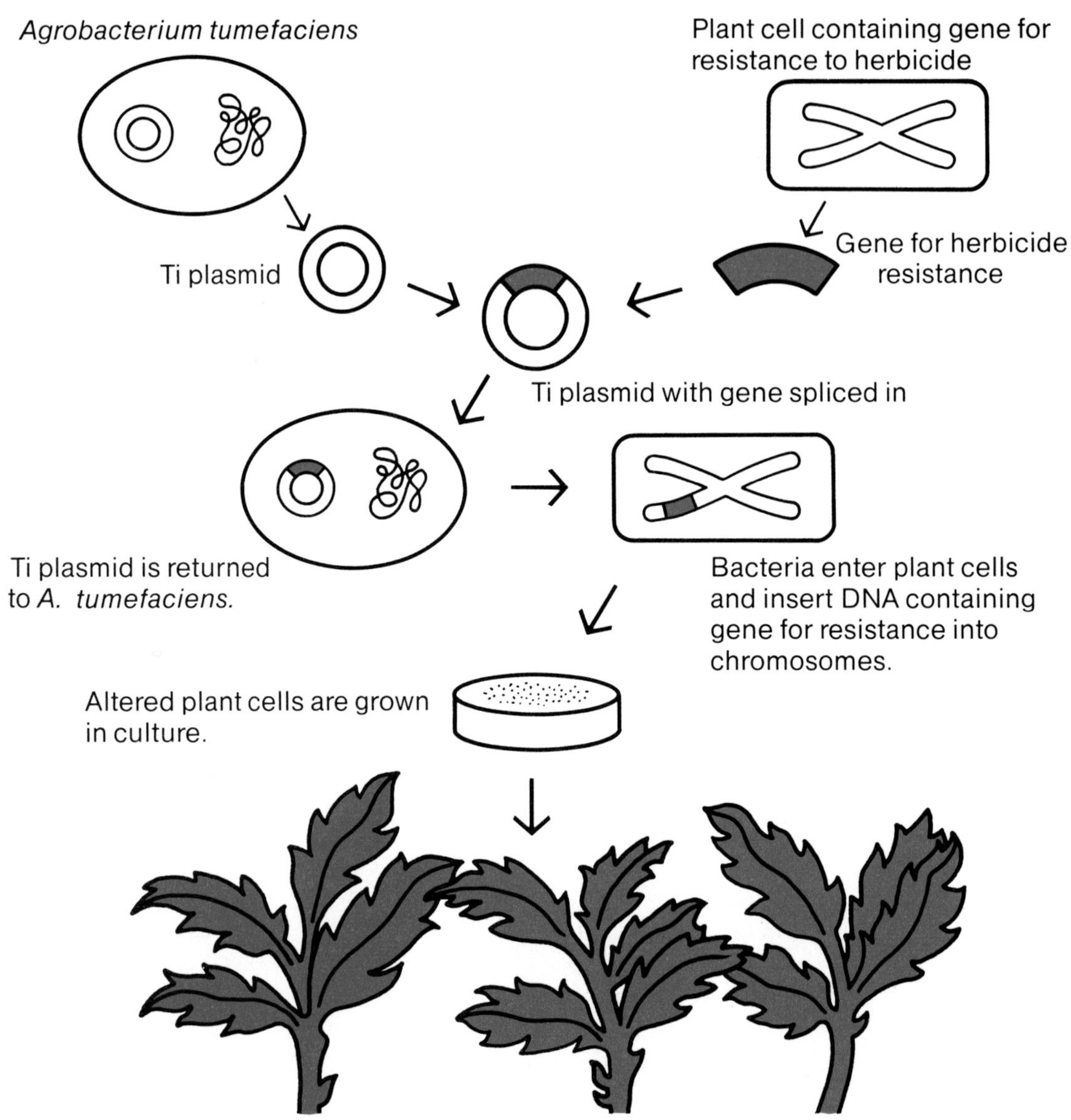

Plants produced from altered cells are resistant to herbicide.

These petunia plants were sprayed with the herbicide Roundup,® which normally kills all green plants. The petunias in the back row survived because they had been made resistant to the herbicide through genetic engineering. The unaltered petunias in front were destroyed by the poison.

IMPROVED PROTEIN QUALITY

Of the 20 amino acids that make up proteins, there are eight that the human body cannot produce but must get from other sources. Known as **essential amino acids**, they are present in animal products such as meat and milk. One or more of them, however, is always lacking in plant proteins. For instance, proteins found in wheat are deficient in the essential amino acid lysine. If lysine could be added to the wheat protein, the nutrition of people who use wheat as a major protein source would be greatly improved.

Storage proteins are the parts of a plant that provide food for people and animals, as well as nourishing sprouting seeds. Researchers are attempting to alter the genes of these proteins so that they code for the production of essential amino acids like lysine. They are studying ways to make small changes in the genetic material of storage proteins that would not disrupt their structure or affect other parts of the plant. Another possibility is the introduction of genes that code for a foreign protein rich in the missing amino acid.

The genetic material for altering plants may be obtained from existing crop plants or wild, uncultivated plants that possess the desired trait. Today plant seeds from all over the world are maintained in gene banks, their genetic material available for transfer into other plants. Traits that are controlled by one gene or only a few will be the first to be introduced. Changes that involve multiple genes, such as increasing growth rates, improving the efficiency of photosynthesis (the use of sunlight to convert carbon dioxide and water to sugars), or conferring the ability to fix nitrogen, will be much more difficult and will take years to accomplish.

The tomato is one of the crop plants that can be grown from a single cell. This ability makes it possible to produce many plants with exactly the same genetic characteristics.

PLANT CELL CULTURE

So far we have concentrated on the hamburger bun in our typical lunch and the changes that biotechnology may bring to the cultivation of wheat and other grain crops. Now let's take a look at those golden, greasy French fries.

The potato and some other species of plants have the potential, under proper conditions of light, nutrition, temperature, and growth-regulation substances, to grow from a single cell into a complete plant with stems, roots, and leaves. By using this method of cell culture, many plants can be obtained from a single plant with desirable characteristics such as superior color or texture, viral resistance, and so on. The new plants can then be planted in fields to mature and reproduce.

Other crop plants capable of being produced by cell culture include asparagus, cabbage, carrots, strawberries, and tomatoes. In the monocot group, rice plants have been successfully grown from a single cell in the laboratory, and work is under way to grow corn and wheat by such means as well. Plant cell culture itself is not an rDNA technique, but it could be combined with genetic engineering. Genetic changes could be made in one cell, which would then be allowed to grow into a new plant with the desired characteristic. Cells from this plant, each containing the altered genetic material, could be used to produce many more plants.

FROST PREVENTION

Frost damage is a recurring fear of farmers in many areas of the United States, and for good reason. Each year, frost causes as much as $1.5 billion dollars worth of damage to potatoes and other food crops. Some of the destruction is brought about by *Pseudomonas syringae*, a bacterial species that occurs naturally on plant leaves. It makes a protein that starts the formation of ice crystals, which can disrupt plant cells. Plants harboring these bacteria are affected by frost at a higher temperature than those without it.

Biotechnology researchers have been able to delete the frost-forming gene in *Ps. syringae*. When the altered bacteria are applied to the leaves of plants, they take the place of the frost-promoting wild types.

In 1985, Advanced Genetic Sciences, Inc., in California was given a license by the Environmental Protection Agency to conduct field tests on strawberry plants that had been treated with the altered bacteria. Greenhouse trials had already been done, and the results were promising. Permission was later withdrawn, however, after a legal suit was brought by those who felt that the release of genetically altered microorganisms might be harmful to the environment. They objected to release despite the fact that mutant strains of *Ps. syringae* lacking the frost-forming protein exist in nature and are not known to do any harm.

After being denied permission to conduct the field test in several different localities, AGS was finally allowed to test its product in Brentwood, California. In April 1987, an AGS worker sprayed a strawberry field with Frostban, the company's name for the two strains of altered bacteria. Preliminary results indicated that the mutant bacteria were effective in preventing frost and did not spread beyond the field on which they were applied.

Release of genetically altered bacteria and other microorganisms into the environment is one of the major controversies of biotechnology. Ecologists and others are afraid of disturbing nature's balance, and biotechnologists must take such concerns into account. As outlined in government regulations, there must be testing for possible harm in experiments moving from laboratory to greenhouse, then to isolated field plots before dispersal in an open ecological system.

It has also been proposed that safeguards be built into microorganisms to be released. These safeguards might be in the form of marker genes that code for a detectable substance to pinpoint the

Scientists at Agracetus, the agricultural branch of Cetus Corporation, examine genetically altered tobacco plants growing at a test site. Field testing is an important part of developing new plants through genetic engineering.

organisms' location. It might also be possible to program the microorganisms genetically for a limited lifespan or to build in sensitivity to an antibiotic, herbicide, or pesticide that could be used to kill them if they spread.

Such safeguards may eventually be considered necessary. Many scientists believe, however, that in testing these products of biotechnology, it is important to sort the potentially risky releases from those that are extremely unlikely to cause any harm.

ANIMAL AGRICULTURE

The beef in the cheeseburger you may have eaten for lunch today did not come from a genetically engineered Angus steer. In the future, it may be possible to alter the DNA of meat animals. The major contribution of biotechnology today, however, is to help produce healthier and more numerous animals by providing new ways to prevent and control diseases and to promote better nutrition and growth.

MONOCLONAL ANTIBODIES

Monoclonal antibodies are useful in the diagnosis and treatment of disease of animals, just as they are in human health care. The first monoclonal antibody to be licensed for prevention of disease was a product for cattle. It is Genecol™99, effective against the bacterial form of scours, an infection of the intestinal tract of newborn animals.

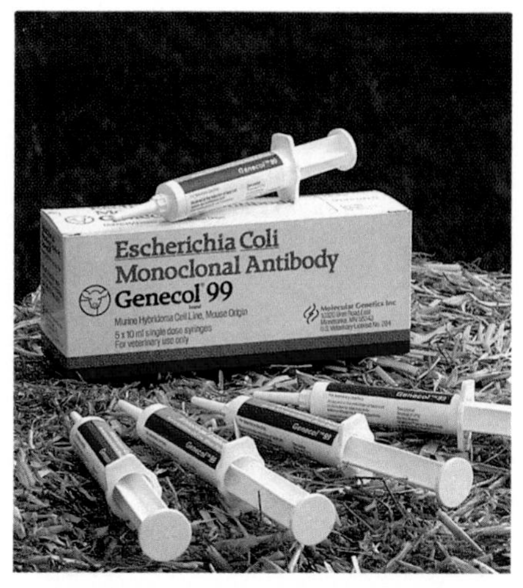

Genecol™99 being administered to a calf. This product, developed by Molecular Genetics, Inc., was the first monoclonal antibody licensed for use in disease prevention.

Scours causes severe diarrhea and dehydration, and often results in death. As many as a million calves in the United States may die from it each year, with an estimated loss of $250 million to farmers. Millions more calves need intensive treatment in order to survive the disease. The monoclonal antibody against scours was developed by Molecular Genetics, Inc., along with a diagnostic kit to detect the presence of the disease in a herd.

Diagnostic monoclonal antibodies are being developed for a variety of diseases of horses and cattle, as well as parvovirus, rotavirus, and heartworm in dogs.

VACCINES

Safe, effective, and inexpensive vaccines against diseases of cattle and other animals could have considerable impact on the amount of meat available for a world full of people who need protein. One of the diseases against which a new vaccine is badly needed is foot-and-mouth disease (FMD), which produces sores on the mouth and hooves and severely

Trypanosomiasis is a parasitic disease affecting livestock in many parts of Africa. Researchers hope to use biotechnology to develop a vaccine against it.

weakens the animal. Caused by a virus, it spreads rapidly and can be transmitted to humans.

While FMD does not occur in the United States, it is widespread in South America, the Far East, and Africa, often in areas where animal loss can be least afforded. At present the disease is controlled by slaughtering infected animals. There is a vaccine made from hamster cells, but it is very expensive and must be given yearly. It is therefore not practical for large herds.

There are over 60 different strains of FMD, which makes the development of a vaccine difficult. A genetically engineered vaccine using viral coat proteins is being devised. It should be possible to make a vaccine against all known strains, the use of which could save hundreds of millions of dollars in cattle loss every year.

Another disease of livestock that is serious particularly in the tropics is trypanosomiasis, called sleeping sickness when it occurs in humans. It is caused by a parasite borne by insects, in Africa by the infamous tsetse fly. Over one-third of Africa, much of the land that is not desert, is affected to the point that livestock cannot be raised successfully. The people are deprived of meat, and because they must till their fields by hand, they cannot grow as much food as they could using draft animals.

Several companies are studying the manufacture of a vaccine against trypanosomiasis. If it proves effective, those areas of Africa could be opened for cattle, sheep, and goat breeding. The ultimate answer to the problem might be the introduction of breeds resistant to the disease. A solution far in the future could be the genetic alteration of livestock by transfer of genes from resistant animals.

Other vaccines being developed for use in animals include those against bacterial diseases such as swine dysentery and pasteurellosis, a respiratory disease of cattle, sheep, and hogs. There are also vaccines against viral diseases, including pseudorabies in hogs, Newcastle's disease in chickens, and rabies; against protozoan-caused coccidiosis in chickens; and against parasites such as hookworms, tapeworms and liver flukes.

A microbiologist at the U.S. Department of Agriculture prepares to inject chicks with monoclonal antibodies against organisms causing coccidiosis. If the antibodies prove to be effective, they will be used in developing a vaccine against the disease.

GROWTH HORMONES

Administering bovine growth hormone to beef cattle to increase weight gain and to dairy herds to raise milk output is under study. Treatment with hormones, which could be produced in quantity by rDNA techniques, would have to be carefully monitored to ensure that no traces of the hormones remain in the product to reach the consumer.

The advisability of treating animals with these substances is a subject of much debate. Some people are concerned that the hormones might impair the health of animals. Others question the wisdom of increasing milk production when there is already an overabundance of milk. Those in favor of growth hormone use, however, claim that it would serve primarily to lower the cost of milk production rather than increasing the supply of milk.

Growth hormone might also aid in animal reproduction. One early study

Administering growth hormones to dairy cattle could increase the already overabundant supply of milk in the United States.

suggests that treating female hogs with hog growth hormone before they give birth may raise the survival rate of piglets. Currently one to three may be lost for every ten born.

FOOD INGREDIENTS AND ADDITIVES

RENNIN

The slice of cheddar cheese on your cheeseburger might be the result of biotechnology. Substances made by various species of bacteria are responsible for the characteristics of the different kinds of cheeses. Cheesemakers would like more stable, dependable strains of bacteria that consistently produce the desired flavor, aroma, and texture, and are resistant to attack by bacterial viruses. Strains that meet these requirements are now being developed by rDNA techniques.

A basic step in making cheese, the coagulation of milk, is accomplished with an enzyme called **rennin**. This enzyme is usually obtained from the stomachs of young calves, one of its few natural sources. The supply of rennin is limited, especially in the United States, where veal is not a popular meat. The uncertainty of the rennin supply leads cheese processors to pay high prices for it (sometimes over $1,000 a pound), import cheeses, or use substitutes. Some of these substitutes are enzymes from microorganisms, but they do not produce totally satisfactory results.

Once again, biotechnology has come to the rescue. The gene for rennin has been cloned in bacteria and yeasts. One company, Genex, is having trials conducted on a rennin it has made using genetically engineered *E. coli*.

Phenylalanine, one of the components of the sweetener NutraSweet,® is produced inside these large bioreactors in a Genex plant in Kentucky.

SWEETENERS

You probably washed down your burger and fries with a soft drink. If it is the regular kind, there is a very good chance that it is at least in part sweetened with a product of enzyme technology, high fructose corn syrup. (Read the label on the next bottle of cola you see.) HFCS is made by treating cornstarch with a series of three enzymes to convert sucrose molecules to fructose, a sugar that is twice as sweet and thus cheaper to use. These enzymes come from bacteria. Engineering bacteria to produce more enzymes with improved properties could considerably reduce the cost of HFCS.

If you are drinking a diet soda, it may be sweetened with **aspartame**, better known by the brand name NutraSweet.® Aspartame is a peptide composed of a molecule of the amino acid phenylalanine linked to a molecule of aspartic acid. Besides soft drinks, NutraSweet is becoming increasingly popular to sweeten such products as soft drink mixes, dessert mixes, and chewing gum. At present, the amino acids are made separately by bacteria, then put together. Ways are being investigated to make the entire peptide in engineered microorganisms.

Aspartic acid and phenylalanine are not the only amino acids produced with microbial enzymes. Some amino acids are presently made so inexpensively that there is no real need for rDNA methods, but the techniques have been applied in Japan to increase production of glutamic acid, used for the food additive monosodium glutamate (MSG). The cost to produce lysine and tryptophan, two amino acids used widely as animal feed additives, might be reduced by the use of genetically engineered microorganisms.

If you wanted to supplement your meal with a vitamin capsule, it will probably come as no surprise that many vitamins are made by microorganisms. Here again, production could be improved and made more economical by rDNA techniques.

SINGLE-CELL PROTEIN

Yeasts and bacteria not only produce food supplements but can also be grown as food themselves. Their cells contain a high percentage of protein. **Single-cell protein** (SCP), the name for microorganisms like bacteria grown for food, has been produced since the 1920s for consumption by both animals and humans. Besides protein, SCP contains carbohydrates, fats, vitamins, minerals, and nucleic acids. Unlike plant crops, its growth is not dependent on temperature, sunlight, or rainfall. SCP could therefore prove to be a valuable commodity in regions where chronic protein shortages from the quantity and quality of the food grown affect physical as well as mental development.

Cells for SCP are grown in large amounts, harvested, and dried. The resulting tasteless flour can be used as a protein supplement for animal feed or human food. (If the idea is unappealing, remember how many of the foods we eat every day, including cheese, yogurt, and bread, are made using microorganisms that remain in the product.) In human food, SCP is used to add functional qualities such as whipping ability. Before consumption by humans, it must undergo treatment to reduce the high nucleic acid content. SCP for animal feed requires less processing because animals can degrade larger amounts of nucleic acids.

SCP can be grown on such diverse substances as methane, wood chips, and wastes from cheesemaking and food processing, thus changing inedible matter into food. Yeasts for animal fodder are grown in several countries on sulfite waste from paper mills, turning a potential major pollutant into edible protein. In England, SCP is produced on methanol for use as animal fodder. Called Pruteen,® it has a crude protein content of 80 percent and is rich in vitamins.

These rice farmers in India, like millions of people in other developing countries, have diets that are lacking in protein. Single-cell protein and other products of biotechnology may eventually improve nutrition around the world.

At present, a major factor working against the widespread use of SCP is its cost. It is more expensive to produce than other protein supplements such as soybean meal and fish meal. Researchers are trying to alter the genetic makeup of SCP microorganisms by rDNA methods to raise their nutritional content, to permit growth on a wider variety of waste products, and to increase yield. Such improvements could reduce costs and make SCP competitive with other protein supplements.

Increased production of single-cell protein is only one of the ways in which biotechnology may eventually affect the world's supply of food. In the future, the use of genetic engineering techniques may help to provide better nutrition for billions of people.

The Kennecott open-pit copper mine near Salt Lake City, Utah. Microbes have played a role in copper mining for many centuries.

5

RESOURCES AND INDUSTRY: BIOTECHNOLOGY IN THE ENVIRONMENT

Picture the workings of an enormous copper mine: giant shovels scooping tons of earth from a hillside; conveyer belts; huge trucks; piles of spent ore; dusty miners in hard hats. Now imagine billions of microscopic miners at work deep beneath the ground, converting insoluble minerals into soluble salts that can be washed out and recovered.

Mining metals is only one of many applications in resource recovery and industry that use microorganisms or the enzymes they make. Some of the others are oil recovery, fuel production, and the manufacture of organic chemicals. Microorganisms can also clean up wastes from industrial processes and pollutants from a variety of sources. One of the important future applications of biotechnology will be in improving and expanding this microbial work force.

MINING

Historians believe that microbiological mining may date back to at least 1000 B.C., when it was used by the ancient Romans to recover copper leached by bacteria. It was not until the 1950s, however, that the role of bacteria in the process was conclusively demonstrated.

Several species of bacteria are involved in mining, but the most common is *Thiobacillus ferrooxidans*. Belonging to a class of bacteria known as autotrophs, *T. ferrooxidans* obtains carbon from the carbon dioxide in the air and energy from its action on metals. No sugars or other nutrients need to be supplied. Mining microorganisms live and grow contentedly under conditions of high acidity and high salinity that would kill most bacteria. Besides copper, mining bacteria have a taste for uranium, zinc, nickel, cobalt, and lead.

The system of microbial mining most common today is **dump leaching**, which is used to extract copper from low-grade ores and waste material. In this process, the copper-containing material is transported from an open-pit mine to a dump site, an area often located on a natural

Leaching operations at the Kennecott copper mine. Bacteria occuring naturally in the copper ore cause a chemical reaction that leaches out the valuable metal.

slope and covered with a material such as clay or asphalt to prevent contamination of ground water.

To start the leaching process, acidified water is sprayed, pumped, or flooded onto the surface of the material. The water percolates through the ore and waste, providing an ideal environment for the growth of *T. ferrooxidans*, which occurs naturally in the type of rock that contains copper.

To get the energy they need to live, the bacteria attack metal compounds in the rock, starting a series of chemical reactions that result in the formation of copper sulfate, a soluble salt. The water sprayed on the dump picks up the salt as it filters through the material. After the water is collected, the copper salt is separated from it by chemical means. More than 10 percent of the copper produced in the United States is mined by this leaching method.

In the future, improved technology may make it possible to use tiny microbial miners for leaching copper ore not only in surface dumps but also in underground mines. This method would eliminate transporting tons of ore, crushing and grinding it, and disposing of the waste material. It would reduce soil erosion, as well as the energy use and pollution connected with both mining and smelting, the heating of ore to obtain metal. Bacterial mining could also be used in areas where heavy equipment cannot be taken.

Today bacteria are hard at work mining uranium as well as copper. In Canada, for example, uranium is being extracted by several microbial methods. In one operation, ore was exploded to break it up and then leached on the spot. Canadian miners have pumped water into underground mines to leach out uranium left behind by other mining methods. Leaching is also used to recover uranium from waste dumps.

Mining with microorganisms is not a new method, but recombinant DNA technology may be able to expand its use tremendously. Researchers are now trying to develop new strains of bacteria that could be applied to ore rather than relying on the resident microbial population. Lodes of ore could be inoculated with bacteria custom-made for the demands of each individual site. Bacteria might be engineered that are more tolerant to acid and toxic metals and better able to withstand the higher temperatures that would be encountered in deep mining operations. It might even be possible to alter bacteria to mine such valuable metals as silver, gold, platinum, mercury, and cadmium.

OIL RECOVERY

The search is on for microorganisms to be used for improved oil recovery. It is estimated that traditional methods of extracting oil leave 50 percent of underground reserves untouched. Much of the remaining oil is too thick to pump or is trapped in rock or tar sands.

There are several possible ways to enlist the aid of microorganisms to get more oil out of the ground. One is to inject bacteria along with nutrients into wells, where they would produce detergents and emulsifiers to free oil from rock or sands. Bacteria could also form gums and gas that would force out the oil. Researchers would like to find or modify bacteria that feed on the oil itself but only on the less valuable components, leaving the rest to be recovered for human use.

An alternative to sending bacteria into wells would be to engineer them to make large quantities of substances such as xanthan gum, which have been found to be effective for oil recovery.

FUELS

Besides helping to recover energy resources, microorganisms can produce fuels such as methane, methanol, and ethyl alcohol (better known as ethanol). Microbial fuel production often utilizes waste products, thus accomplishing a dual purpose. The costs to carry out these processes must be lowered, however, before they can compete with traditional methods. Recombinant DNA technology may make such fuels less expensive to produce.

INDUSTRIAL CHEMICALS

In addition to its use as a fuel, ethanol is a very important industrial solvent, second in use only to water. It is necessary for making a wide variety of products, including other solvents, pharmaceuticals, lubricants, cosmetics, and dyes.

Part of the ethanol supply for industry is made synthetically, but the bulk of it comes from growing yeasts on sugar beets, molasses, and grains such as corn. Yeasts can also be grown on such wastes as wood pulp and leftovers from food processing. Using these materials involves the breakdown of cellulose (long chains of glucose molecules) by cellulases, enzymes produced by fungi. The process could be made more economical and efficient by bacteria genetically altered to make large amounts of cellulases. If more waste products could be used in this way, more grain could be spared for food. Ethanol made from waste products could also be a valuable energy source in developing countries.

One of the largest expenses in making ethanol is distilling off the product. One manufacturer is reducing this cost by using yeasts that have been genetically altered to withstand the high temperatures used to distill off the product as it is secreted by the yeast cells. Ethanol manufacture is a perfect candidate for the continuous flow bioprocess. Yeast cells are immobilized on a column, nutrients are pumped into one end, and the alcohol is recovered at the other.

Bioprocesses can often replace traditional processes for making other organic chemicals, such as alkene oxides for plastics, acetone, butanol, and isopropyl alcohol. Where this is possible, there can be a number of advantages over nonbiological methods. Microorganisms and their enzymes do not need extremes of heat and cold, polluting solvents, or toxic metal catalysts that chemical processes require. Instead, they respond to the mild conditions favored by most living things. There are fewer byproducts because most commodities are made directly, without intermediate steps. Even the manufacturing facilities can be cheaper to build and run.

This large oil spill in the Gulf of Mexico was caused by release of oil from a offshore well in 1979. In the future, genetically engineered microorganisms may be used to clean up oil spills and other kinds of pollution.

POLLUTION CONTROL

The earth is polluted with millions of tons of waste products that will not go away. Industrial processes create a great deal of pollution. Undegraded pesticides and herbicides linger in the soil and water. Oil spills leave their mark on beaches and the ocean floor.

Microorganisms can help clean up the mess. They already do a massive amount of waste and garbage degrading without being asked. They are also employed in controlled systems such as sewage treatment plants, where they break down biological wastes, grease, plant matter, and other materials that we wash down the sewer.

Microorganisms have been found that can degrade a variety of substances. Through the process of natural selection, bacteria exposed to a particular substance develop the ability to live in its presence and eventually degrade it. With genetic engineering, it is now possible to speed up the process and tailor the organisms to specific needs. Bacteria that grow in the laboratory on a given substance can be singled out and their genes introduced into other organisms so that they too will have the ability to degrade the substance.

TOXIC WASTES

It is possible that a microorganism can be engineered to degrade just about anything. Researchers are developing bacteria capable of degrading cyanide, some dioxins, and polychlorinated biphenyls, or PCBs, a particularly long-lasting and toxic class of chemicals. Strains of *Pseudomonas* have been found that degrade the herbicide 2,4-D, and a modified *Pseudomonas* can break down the herbicide 2,4,5-T. (Both of these organic compounds are components of the deadly Agent Orange.) It might be possible to genetically alter a microbial strain or a mixture of microorganisms to be added to toxic waste dumps to degrade all the offending substances.

Many problems remain to be solved before we can make full use of pollution-degrading microorganisms. Care must be taken that they do not produce harmful byproducts in the process. Information on such matters as how long the microorganisms live, where they go, and how they interrelate with other organisms must be collected before this aspect of biotechnology can be fully developed.

CONCENTRATION OF METALS

Heavy metals are often present in the waste water produced by industrial processes. Such metals as copper, nickel, gold, silver, uranium, zinc, and cadmium can end up in lakes and streams, and eventually public water supplies unless they are removed from waste water. Some bacteria and yeasts are capable of concentrating certain dissolved metal, either accumulating them on their walls or taking them up inside the cells. This ability is useful in retrieving valuable metals for reuse and at the same time preventing water pollution.

One possible method to achieve these goals is to immobilize the microorganisms and pass the waste water over them. In the case of toxic metals, metal-concentrating proteins produced by genetically altered bacteria could be used instead. Genetic engineering could give microorganisms the ability to concentrate more kinds of metals and in larger quantities.

Biotechnology is capable of fighting industrial wastes in two ways. One is the actual cleanup of wastes. Another is the replacement of current methods with biological processes that use less energy, do not produce toxic wastes, and use fewer nonrenewable resources. Many of these processes are expensive now, but as they become more efficient, there should be a decrease in operating costs.

OIL SPILLS

Oil spills from tankers and offshore oil rigs do untold damage to ocean and shore life, damage that could be reduced by microorganisms. A number of bacteria exist in nature that can feed on various components of petroleum. Such bacteria have been used at sites of oil spills and also in hazardous waste dumps, where they break up coal tar. During the renovation of the ocean liner *Queen Mary*, bacteria were used to clean oil from bilge water to eliminate the risk of acetylene torches setting off an explosion. To improve the operation of such microbial cleaning crews, genes from oil-loving species might be used to create more efficient bacterial strains.

CLOGGED DRAINS

To solve a problem on a smaller scale, bacteria are being engineered to decompose grease that results from the processing of meat and poultry. This grease can build up, clogging drains and causing odors, or create problems in waste treatment plants.

Some applications are closer to home. Proto™ is a drain cleaner developed by a biotech company, Genex. Composed of enzymes made by the soil bacteria *B. subtilis*, it dissolves hair clogs. Products such as these could replace current drain cleaners that employ such caustic substances as lye and generate a considerable amount of heat. The company plans an entire line of similar products. Other enzymes, many also from *B. subtilis*, are now added to laundry detergents to digest stains.

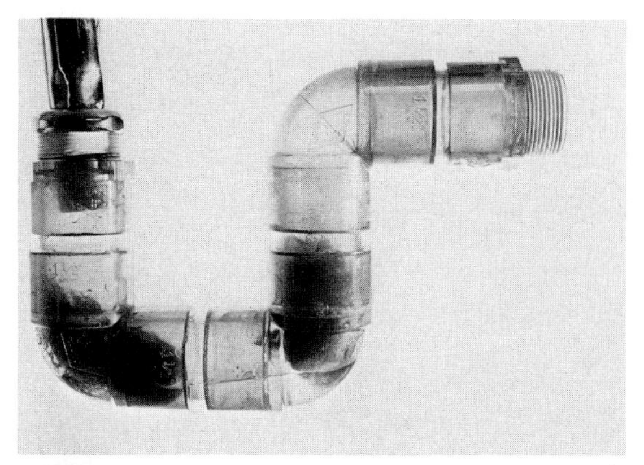

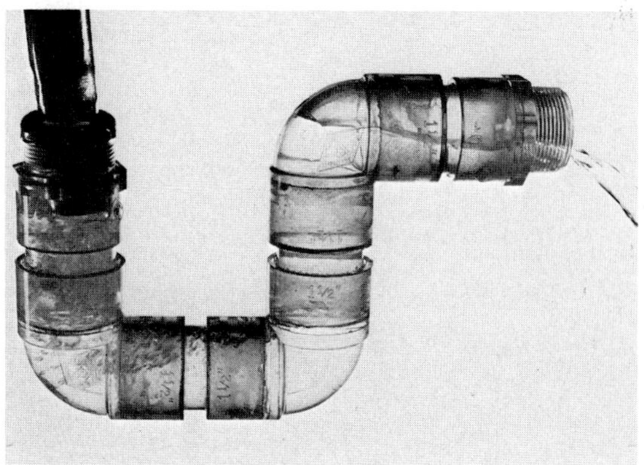

You may soon be able to use a product of the new biotechnology in your kitchen sink. Proto,™ a drain cleaner developed by Genex, uses enzymes made by a species of bacteria. These photographs show Proto at work on a clogged drain.

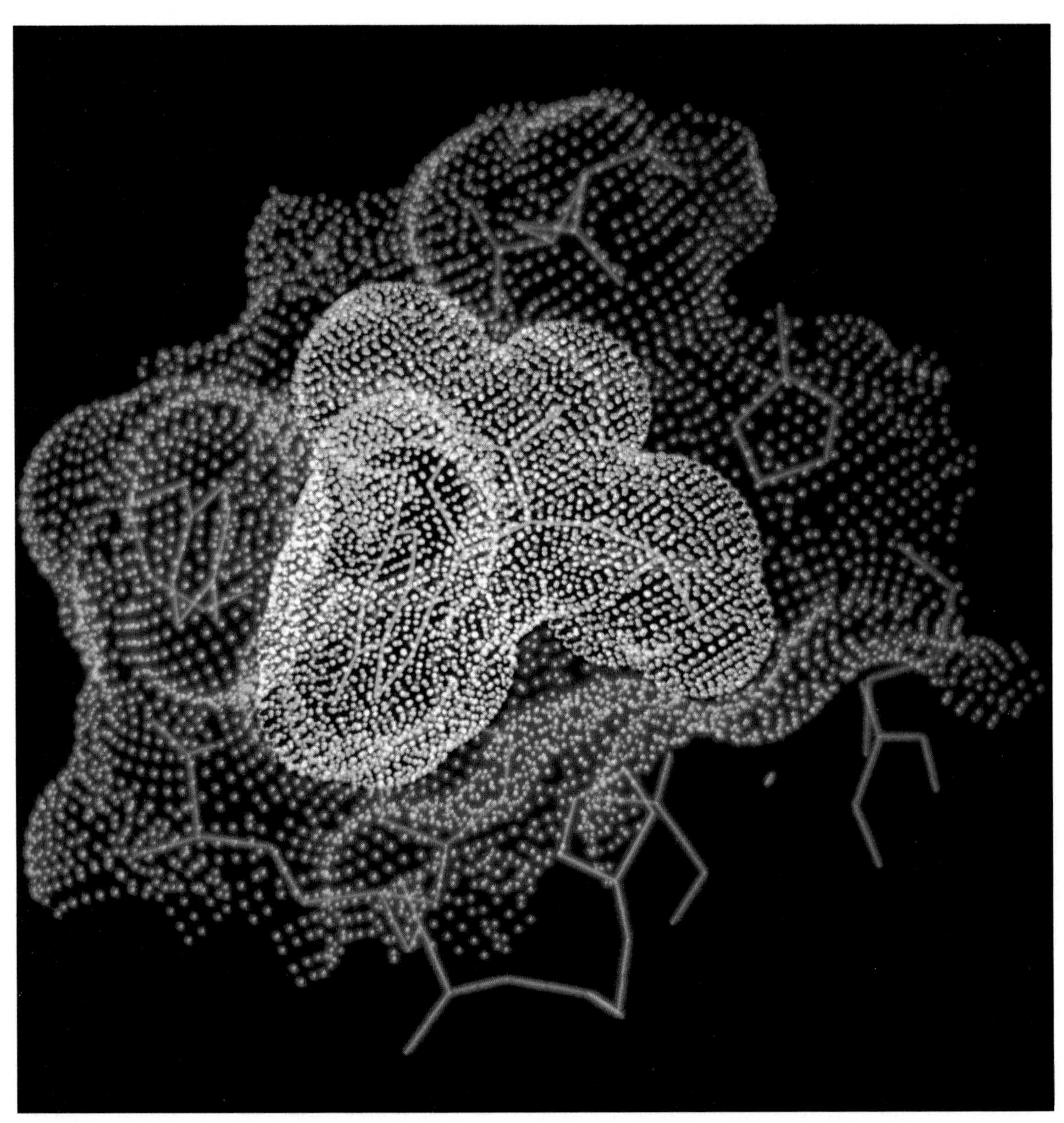

Protein engineering, one of the new areas of research in biotechnology, makes use of computer graphics to design and modify proteins. This computer image shows an enzyme, ribonuclease T1 (blue), binding to the substance on which it acts (red).

THE FUTURE OF BIOTECHNOLOGY

A decade or so ago, the term biotechnology did not even exist. Yet today, some of the most fantastic early predictions about what biotechnology could do have become fact. There can be little doubt that proposed products and processes considered impossible today will be commonplace 10 years from now.

Three areas that sound like science fiction today are gene therapy, bioelectronics, and protein engineering. All three are currently subjects of research, although their full potential lies in the future.

GENE THERAPY

Gene therapy would make use of our knowledge of genetics and recombinant DNA to correct defects in genetic material itself. The first human gene therapy will probably be treatment of an enzyme or hormone deficiency caused by a defective gene. It will involve the introduction of a normal gene to replace the malfunctioning one. The method for administering gene therapy under study is to remove bone marrow from the patient, add the corrective gene, and return the marrow. The new gene would then direct the synthesis of the missing protein.

As with plants, the first successful transfers will probably be of a single gene. The conditions being considered for treatment with gene therapy are very rare, very serious diseases, ones that may affect only 10 to 1,000 people and for which there is no treatment. Two such diseases are adenosine deaminase deficiency and Lesch-Nychan syndrome, also an enzyme deficiency. Among more widespread disorders, two possible candidates for treatment are thalassemia and sickle cell anemia, both blood diseases caused by defects in genes coding for hemoglobin.

Replacing a gene that cures a disease in an individual but is not passed on to offspring is no different from any other type of medical treatment. Altering genes so they can be transmitted to future generations is not only technically impossible at present but is also surrounded by serious ethical questions.

During the 1970s, advances in computer technology made it possible to replace the large integrated circuit on the right with the much smaller unit on the left. In the future, the use of protein-based biochips may lead to equally significant developments in computer science.

BIOELECTRONICS

The field of bioelectronics, the use of protein molecules in electronic devices, is in its infancy. It is based on the fact that polypeptides, or strings of amino acids, are capable of conducting electricity. There is work being done in two areas, biosensors and biochips.

Biosensors, which are monitors that use enzymes, monoclonal antibodies, or other proteins, have been in use for several years to test for the presence of organic compounds. There are problems associated with them, including expense and sensitivity to temperature. Improved biosensors could test air and water quality and detect hazardous substances in factories. In medicine, sensors might monitor blood-sugar and insulin levels for more effective treatment of diabetes, as well as antigens and serum protein levels.

In the future, protein-based biochips might replace silicon chips for certain applications. Computers using biochips would be smarter, faster, more energy-efficient, possibly even more reliable. Some potential uses for computers with biochips are implants in the body to deliver precise amounts of drugs, to control artificial limbs, and to regulate heart rate. There might even be implants that would aid the blind and deaf. Bioelectronics is a promising area of research, but the technical problems are immense, and success is a long way in the future.

PROTEIN ENGINEERING

Protein engineering may be biotechnology's next frontier. Founded upon genetic engineering and other technologies such as computer graphics and X-ray crystallography, it is the designing and modifying of enzymes and other proteins.

Two ways of engineering proteins are altering the structure of the gene that codes for a protein and building a synthetic gene. Using these techniques, proteins might be made more stable, more efficient, useful under a wider range of conditions, or able to perform additional functions.

In medicine, for instance, proteins might be altered so that, when given as therapeutic drugs, there is less destruction of them in the digestive system or in the blood. This would make smaller doses possible. There are also potential applications for engineered proteins in diagnosing disease and as enzymes for industrial use.

HELP FOR THE FUTURE

Biotechnology is a field that is building on the past to revolutionize the future. It has the potential to be a major force in human society, influencing the way in which we treat the sick, produce food, manufacture goods, and recover resources. As various technologies in these areas are further developed, costs will drop, making them more competitive with methods now in use.

Biotechnology cannot solve all the ills of humankind, and not all new developments will find applications—some things are still better done using more traditional methods. As we have seen, there are also possible dangers involved in some rDNA techniques. Few advances in science or technology, however, are made without risk. It is the task of researchers, commercial users of the various techniques, and regulatory bodies to make sure that no unnecessary risks are taken and to win and keep the trust of the public.

The most important goal for all concerned is to explore thoroughly the ways in which biotechnology can be used to improve the quality of human life. The potential is there, and we need all the help we can get.

A fermenter in a Cetus plant manufacturing therapeutic products. The color coding on the pipes identifies the many different materials supplied to the fermenter and removed from it during the production process.

GLOSSARY

amino (uh-MEE-noh) acids—organic acids that are the building blocks of proteins. The order of amino acids in a specific protein is determined by the order of bases in the gene directing the production of that protein.

antibodies—protein molecules produced by the immune system in response to a foreign substance such as a bacterial cell

antigen (ANT-ih-jehn)—a foreign substance that stimulates the production of antibodies when it enters the body

antigenic (ant-ih-JEHN-ik)—causing an immune response

aspartame (AS-puhr-tame)—a sweetnener made by combining the amino acids phenylalanine and aspartic acid. Sold under the brand name Nutrasweet,® aspartame is produced by bacteria.

bacteria (singular, bacterium)—single-celled organisms often used as hosts in genetic engineering

bioprocess—any operation or system that uses living cells or some component of cells to make a product

bioprocess engineering—planning and setting up equipment and facilities to make the product of a bioprocess on an industrial scale

cell fusion—the merging of two different types of cells into a single hybrid cell

chromosomes (KRO-muh-somes)—thread-like structures in a cell nucleus that are made up primarily of DNA

codon (KO-dahn)—a group of three nitrogen bases that codes for a specific amino acid or for different steps in the formation of amino acid chains

complementary pairs—combinations of the bases thymine and adenine or cytosine and guanine that make up the rungs in the DNA ladder

copy DNA (cDNA)—DNA made by a DNA synthesizer or by using mRNA as a template. Copy or complementary DNA is often spliced into the DNA of other organisms in genetic engineering.

deoxyribonucleic (dee-AK-see-ri-bow-nyu-*klee*-ik) acid (DNA)—the chemical substance in the cells of living things that directs the production of proteins and contains genetic information passed on to new cells and new organisms

DNA synthesizer (gene machine)—an instrument that puts together chains of nucleotides to form gene sequences. The gene machine can be used to make cDNA for genetic engineering.

dump leaching—the process of extracting copper and other metals from low-grade ore and waste materials by the action of a bacterial species, *T. ferrooxidans*

enzymes (EN-zimes)—proteins that speed up chemical reactions but are not themselves changed by them. Certain enzymes are used in rDNA work to cut and join pieces of genetic material.

essential amino acids—the eight amino acids that the human body cannot produce but must get from other sources such as animal products

genes (JEANS)—regions of DNA on a chromosome, each of which contains the genetic code for the production of a specific protein

genetic code—the group of 64 codons that contains the instructions for forming amino acid chains to produce proteins

hybridomas (hi-brid-OH-muhs)—hybrid cells formed by the fusion of two different cells

immune response—the response of the immune system to foreign substances entering the body. The production of antibodies and interferons is part of the immune response.

interferons (in-tuhr-FIHR-ahns)—proteins that control the body's response to viruses and cancer cells by alerting the immune system to the presence of the disease-causing agents

monoclonal antibodies—antibodies produced by hybrid cells exposed to an antigen. Monoclonal antibodies are identical to each other because they are made by clones of a single cell.

nitrogen bases—chemical compounds that make up the "rungs" in the DNA ladder. The four bases are adenine, guanine, thymine, and cytosine.

nitrogen fixation *(nif)* **system**—the combination of genes that allows some bacteria to change nitrogen into a form that can be used by plants. Researchers hope to splice *nif* genes into plants so that they can fix nitrogen themselves

nucleotide (NYU-klee-uh-tide)—the unit of DNA or RNA formed by a nitrogen base joined with a sugar and a phosphate

plasmids (PLAZ-mihds)—rings of DNA in bacteria that are separate from the single chromosome

polynucleotides—chains of nucleotides

protein—a complex chemical substance, consisting of amino acids, that makes up the structure of cells and performs many functions within them

purines (PYUR-eens)—one of the two groups of nitrogen bases in DNA and RNA. Purines contain atoms that are joined to form two rings.

pyramidines (pie-RIM-eh-deens)—one of the two groups of nitrogen bases in DNA and RNA. Pyrimidines contain atoms that are joined to form one ring.

recombinant DNA (rDNA) technology—methods and techniques used to remove DNA from one organism and combine it with the DNA of another organism. Following the instructions on the foreign DNA, organisms such as bacteria and yeasts can produce substances that they are normally unable to make.

rennin (REN-uhn)—an enzyme that coagulates milk; used in the making of cheese

replication—the process by which DNA makes copies of itself in preparation for cell division

restriction endonucleases (en-doh-NYU-klee-ays-es)—enzymes used to cut a gene from the surrounding DNA

ribonucleic (ri-bo-nyu-KLEE-ik) acid—a chemical substance made in the cell nucleus using DNA as a template, or pattern. **Messenger RNA (mRNA)** carries the genetic message of DNA into the cytoplasm, where it is used in the production of proteins. Another form of RNA, **transfer RNA (tRNA)**, also plays a role in protein production.

ribosome (RI-buh-sohm)—one of the small bodies in the cell cytoplasm that serve as centers of protein production

single-cell protein (SCP)—microorganisms such as bacteria grown as food

streptokinase (strep-toe-KI-nase)—an enzyme produced by *Streptomyces* bacteria that dissolves blood clots

subunit vaccines—vaccines consisting only of the part of a virus that causes an immune response

Ti plasmid—a plasmid in the bacterial species *Agrobacterium tumefaciens* that causes the crown gall tumor in some plants. The Ti plasmid is often used as a vector in the genetic engineering of plants.

tissue plasmogen (PLAS-muh-jehn) activator (t-PA)—a human enzyme that dissolves blood clots. T-PA can be produced by genetically engineered bacteria and yeasts.

transformation—the process of inserting recombinant DNA into host cells

urokinase (yur-oh-KI-nase)—a human enzyme that dissolves blood clots

vaccinia (vak-SIN-ee-uh) virus—the virus used in making vaccine against smallpox

vector (VEK-tuhr)—the carrier that receives a foreign gene and inserts it into a host organism. Bacterial plasmids are the most commonly used vectors in rDNA technology.

viruses—tiny organisms consisting of a protein coat and an inner core of genetic material. Viruses live in the cells of other organisms and often cause disease.

yeasts—single-celled fungi, sometimes used as hosts in genetic engineering

ACKNOWLEDGMENTS The photographs and drawings in this book are reproduced through the courtesy of: pp. 6, 8, 31, 32 (bottom), Interferon Sciences, Inc.; pp. 10, 37, 75, 85, Genex Corporation; pp. 12, 14, 15, 17, 18, 20, 24, 25, 26, 66, Laura Westlund; p. 19 (left) David Peters, University of California, San Francisco; p. 19 (right), Stanford University School of Medicine; pp. 21, 23, 28, 29 (right), 38, 49, 50, 56, 57, 70, 90, Cetus Corporation; pp. 29 (left), 42, Genentech, Inc.; p. 32 (top), Hybritech Inc.; pp. 34, 55, Monoclonal Antibodies, Inc.; p. 40, Minneapolis *Star and Tribune*; pp. 44, 72, 77, United Nations; p. 47, Eli Lilly and Company; p. 51, Library of Congress; p. 53 (left), National Library of Medicine; p. 53 (right), Phil Porter; p. 58, Banco Interamericano de Desarrollo; pp. 61 (left), 62, 64, 68, 73, U. S. Department of Agriculture; pp. 61 (right), 66, 67, 74, 94, Monsanto Company; p. 71, Molecular Genetics, Inc.; pp. 78, 80, Kennecott Copper; p. 83, American Petroleum Institute; p. 86, Drs. Paul Axelsen and Frank Prendergast, Mayo Graduate School of Medicine, Department of Biochemistry and Molecular Biology; p. 88, Control Data Corporation

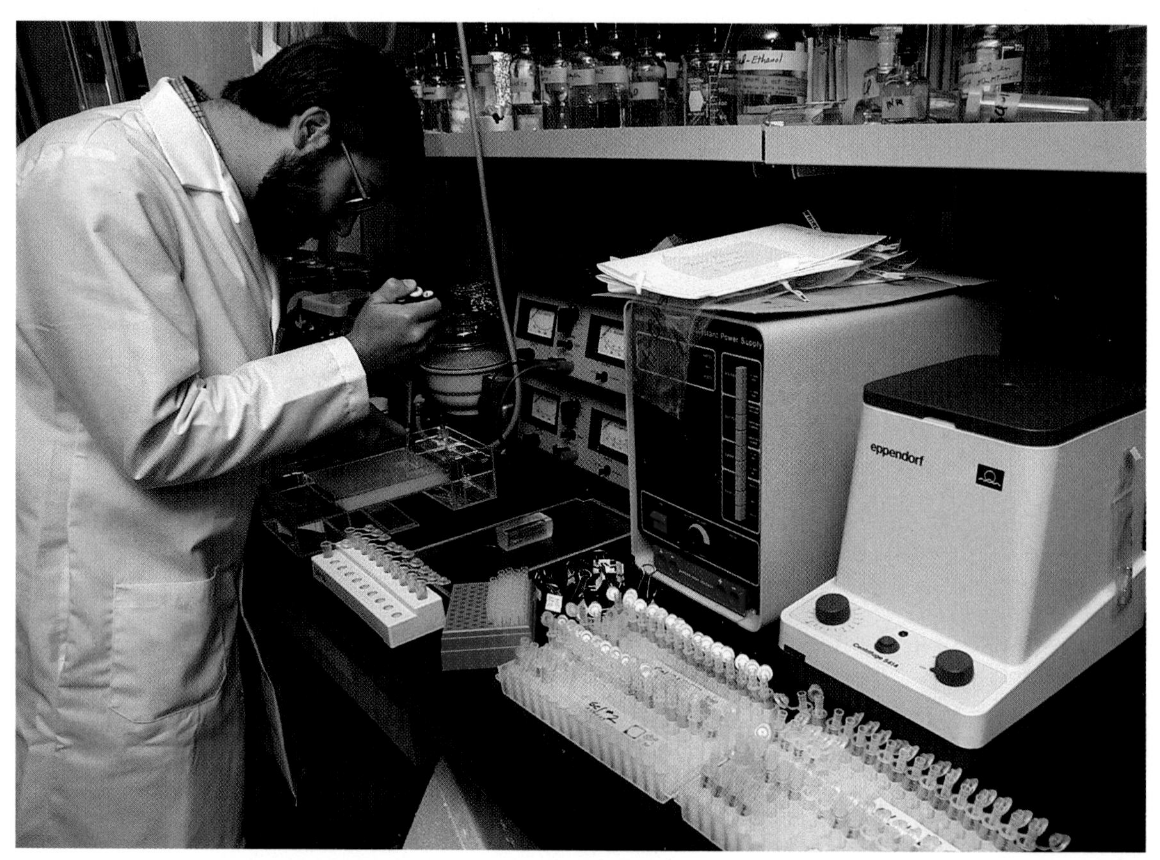

Using a pipette and tiny test tubes, a scientist at Monsanto mixes bacterial plasmids with restriction enzymes that open the plasmids at specific sites. Genes cut from the cells of other organisms by the same enzymes can then be spliced into the plasmids, forming recombinant DNA.

INDEX

j Gross, Cynthia S.
620.8 The New biotechnology.
Gro

Kirtland Public Library
Kirtland, Ohio